T0092967

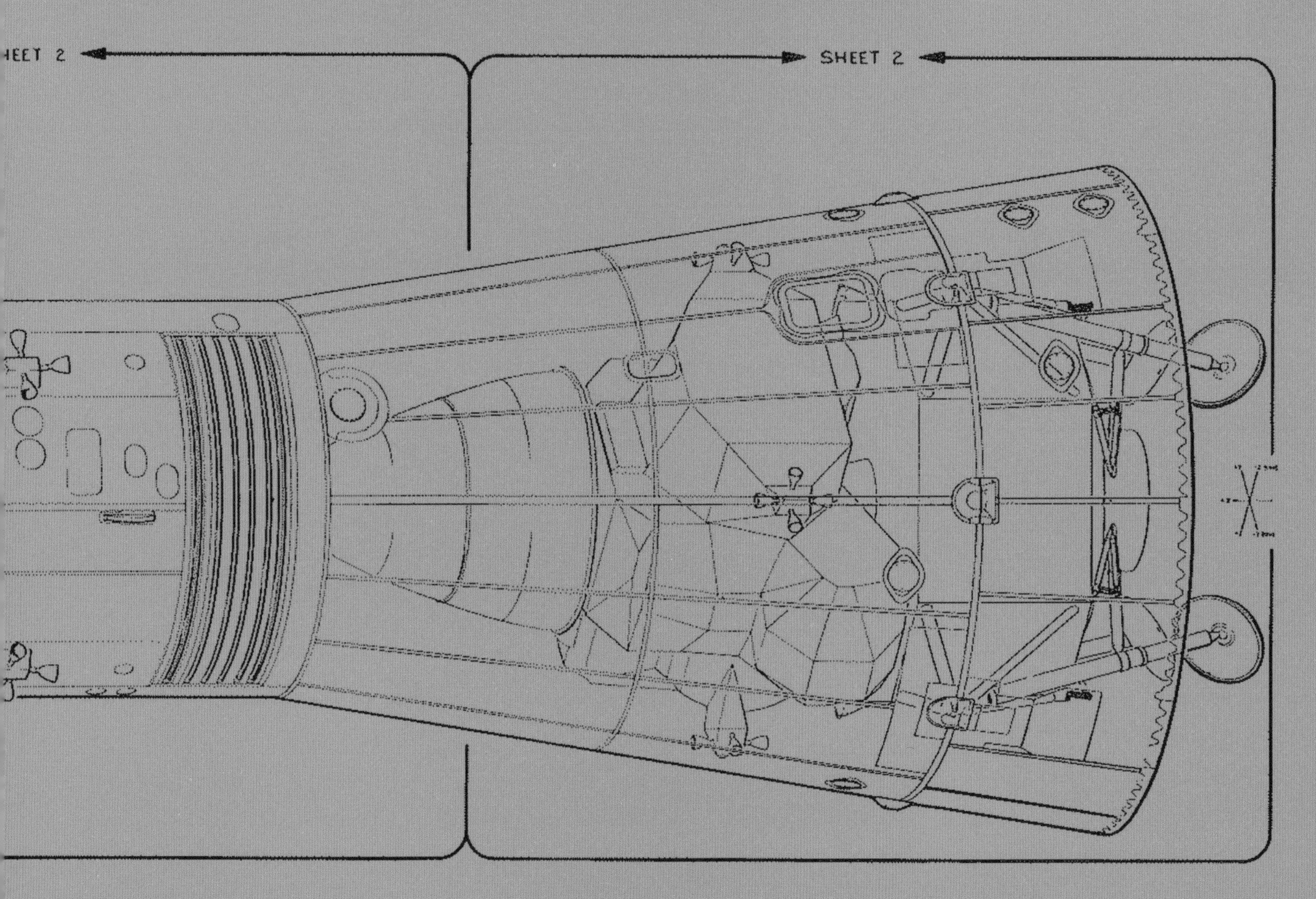
HEET 2
SHEET 2

SPACE HARDWARE

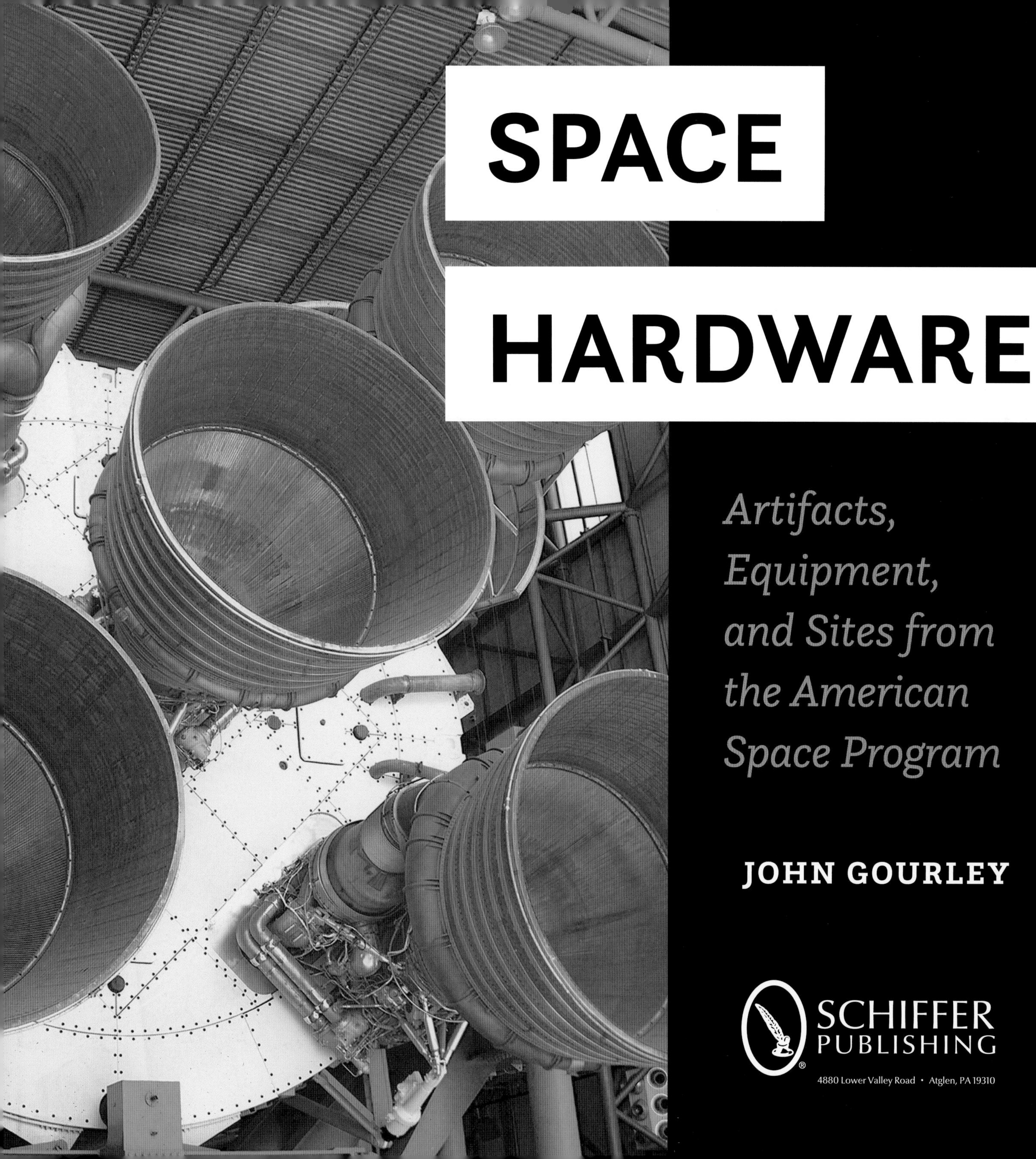

SPACE HARDWARE

Artifacts, Equipment, and Sites from the American Space Program

JOHN GOURLEY

SCHIFFER PUBLISHING®
4880 Lower Valley Road • Atglen, PA 19310

Library of Congress Control Number: 2022932787

Designed by Christopher Bower
Cover design by Molly Shields
Type set in Questa Slab/Garamond Premier Pro

ISBN: 978-0-7643-6528-7
Printed in China

Published by Schiffer Publishing, Ltd.
4880 Lower Valley Road
Atglen, PA 19310
Phone: (610) 593-1777; Fax: (610) 593-2002
Email: Info@schifferbooks.com
Web: www.schifferbooks.com

CONTENTS

Foreword 7
Introduction 8

Chapter 1 The Beginning: Project Mercury 10
Chapter 2 The Gemini Project 24
Chapter 3 Satellites and Space Probes: Civil/Military Uses 40
Chapter 4 Project Apollo 82
Chapter 5 Skylab and ASTP 120
Chapter 6 Space Shuttle and International Space Station 130
Chapter 7 ARES I-X, Orion, Starliner, Dragon Cargo Vehicle, and Space Launch System 150

DEDICATION

To all of the astronauts, cosmonauts, and space travelers of all nations and to the engineers, designers, and scientists who have pressed ahead with success and continued despite failures, putting into practice realistic, achievable methods within our abilities to satisfy the insatiable natural instinct of every human being to explore not only near-Earth space and the moon, but to send probes to the planets and into the vastness of space outside our solar system. To those who have chronicled past adventures, and who continue doing so today through mediums such as press releases, articles, magazines, and especially photographs.

ACKNOWLEDGMENTS

The author expresses a special recognition and gratitude to Dennis Jenkins, a longtime friend as well as a fellow aviation and space enthusiast. Dennis has provided me with data, photos, and guidance over the years, which has always been, continues to be, and will in the future be greatly and sincerely appreciated. The assistance of many unnamed individuals has been provided in places such as the National Museum of the United States Air Force in Dayton, Ohio; the National Museum of Naval Aviation in Pensacola, Florida; the Space & Rocket Center in Huntsville, Alabama; the Museum of Flight in Seattle, Washington; the Smithsonian National Air & Space Museum in Washington, DC, and the nearby Smithsonian Annex Steven F. Udvar-Hazy Center at Dulles Airport; Cape Canaveral and the adjacent Kennedy Space Center, the Air Force Space Museum, Space History Center and the Astronaut Hall of Fame, and the American Space Museum & Walk of Fame in Titusville, Florida.

FOREWORD

The author has had an obsession following events in space from the time he attended grade school. It was an exciting span of history to witness, as America took on the challenge of sending astronauts into space and out to the moon and returning back to Earth in the 1960s into the early 1970s. Everyone knew someone who either worked for NASA or was a subcontractor supplying components to the Mercury, Gemini, or Apollo manned programs. Others had involvement in the Explorer, Ranger, Mariner, Lunar Surveyor, or Lunar Orbiter robotic missions. Space travel was taught in school. If a launch occurred while classes were in session, the teacher stopped lessons long enough to push into the classroom a wheeled cart, atop which was a television set, so that we could watch live coverage of rocket launches. As a class we were mesmerized as we stared at the black-and-white live video feeds from Cape Canaveral, Florida, while launches of satellites and humans into space happened. The entire nation as well would pause to witness these events. We all had the space-themed lunch boxes and complementary thermos bottles, plastic models you could build at home, and the books and magazines. Our school had science fairs where the dominant theme students would focus on was the moon adventure. We also had the fortune of receiving a few visits from NASA employees, who would quiz us about life in space and test our knowledge of this new topic that everyone had an interest in.

Much else was going on simultaneously, such as the Cold War, the Vietnam War, political events and movements, and there were those who urged that the money be spent on more-pressing domestic programs. Space policy was a source of national pride, especially as we began this new chapter seemingly in second place after the surprising successes of the Soviet Union in sending their Sputnik satellites and cosmonauts in Vostok capsules into space first. America had reasons to stay engaged and do our best to be the first to land on the moon. Plans were created, and they advanced in the form of the Projects Mercury and Gemini series of flights, proving that men could function and live in space for the time it would take to go to the moon, while performing tasks specific to this end and safely return home.

Things such as working in a weightless state; operating equipment unique to the space environment; docking and undocking spacecraft; eating, sleeping, exercising, and performing routine as well as detailed work within set time schedules. None of these capabilities were known in the beginning, and they had to be tried out, recorded, modified if need be, and tested again, until it was deemed safe to move on to the next steps. All of this had to be done within the decade of the 1960s, as laid down by President John F. Kennedy in a speech where he challenged America to go to the moon and back in this decade, after a mere fifteen minutes spent in weightlessness on a suborbital flight by astronaut Alan Shepard in the Mercury capsule Freedom VII. The moon landings are still regarded today as the greatest accomplishment of the United States in the twentieth century.

INTRODUCTION

Contained within these pages is a rare look at space hardware: from Sputnik to SpaceX and everything in between, the objects we have saved for posterity, to marvel at inside museum galleries and out-of-doors sites the world over. It is also a natural desire of human beings to showcase past exploits, and thankfully many relics have been preserved by people with foresight for everyone to view, ponder over, and be proud to have been a part of, whether as a taxpayer or one who actually had a hand in building, designing, or writing specifications, or calculating tolerances and performance values for. Whole images show capsules, satellites, and rockets. Up-close, detailed photos show even more. The actual parts and experiments that made up the entire probe or spacecraft are shown. Some artifacts were spares that would have been used if the original part or craft had not been approved for flight. Some items displayed are not real but were made to represent space hardware by those devoted to this great achievement. Space exploration requires many observations, sampling gear, recorders, and transmitters to send data back to Earth so we here can look at, save, and evaluate the findings for now and the future. Each mission has goals that only handmade, customized robots can investigate; a mass-produced assembly-line type of vehicle will not do. Add to this the harsh, unforgiving environments these satellites and probes that orbit or land on other worlds must be able to function in, and you begin to get a feel for what goes into each and every design.

Manned spacecraft not only must take humans into space reliably but also must carry everything we take for granted in the course of a typical day on Earth. Water. Food. Air. A comfortable working environment. The ability to sleep, exercise options as well as hygiene facilities. Ways and means to conduct routine maintenance as well as repairs of normal wear and tear on equipment, plus to be able to swiftly rectify emergency conditions should they arise. As one example, the atmosphere protects us from meteorites, and most don't even think about this. In space, a meteorite or space junk impacting a manned vehicle could have deadly consequences. In going to the moon, where the gravity is only one-sixth that of Earth, this had a major impact on how tools and vehicles were designed, since weight was a prime factor to keep under control. Even the mighty Saturn V rocket had a limit as to how much weight it could lift into orbit.

The author's passion to photograph space hardware in earnest began with a visit to a cousin and her family living near Washington, DC, in November 1978. With a free day on our schedule, they wanted to take me to the Smithsonian Air & Space Museum on the National Mall. I looked forward to it simply as a fun outing, and we drove into DC on a cold day with intermittent snow showers. I had a 35 mm camera and some film, and very soon after our arrival the past memories of interest in the space topic resurfaced. The galleries were laid out with all sorts of artifacts of the past, some being actual items that were eventually declared excess inventory that were to have been used if needed. The museum's Pioneer space probe (also known as the Pioneer H), as an example, was capable of a mission and would have been the Pioneer 12 mission had the request been granted to launch it. You could actually walk inside the Skylab Orbital Workshop and see what it was like, then imagine being enclosed in such a craft in a weightless state, living and working for extended periods of time. All too quickly, the day was ending and it was time to leave, but despite not really wanting to go, I was determined then to take advantage of every opportunity to add this subject, space hardware, to my growing photographic collection.

Many years since that first outing, my path in life has afforded me many other opportunities to capture and catalog much in all of my photo pursuits, the space subject among them. I have made the most of each and every encounter and spent countless hours afterward making sure I had a working knowledge of exactly what it was that I had taken photographs of. My reasoning was simple. To be able to show others what it was I was so taken with, and also to have a fairly accurate record of my field trips in case other authors asked for my images to use in a book or magazine. With the data accurately annotated (a streamlined process to do in the digital age, but included are some of my scanned wet-film images with handwritten notations as well), it would be less work for fellow enthusiasts to integrate into their drafts. I never had a thought about one day when I would use my own material in a book of my own making!

I am excited to present this detailed examination of many facets of our ventures into space, as illustrated with pictures taken in many locations over many years. The book does not contain a historical account of the space experience, since there exist many such reviews online and in many excellent books one can access. This book will be the perfect complement to any such background documentary. Anyone will soon be able to see the reason why, as the up-close and revealing images show off the equipment used, with some photos highlighted to clearly show the parts that made up the whole platform. Despite recent museum closures and limited access, due to the COVID pandemic, many facilities have taken advantage of this un-planned downtime to undertake gallery renovations as well as adding exhibits and artifacts. This has resulted in enhancing overall visitor experiences. In the meantime, I hope this book can somewhat fill the void with as complete an accounting of what is out there to see as possible, giving inspiration and encouragement to others who have this underlying urge to head out to facilities near and far to take in the full measure of our rich space history. I look forward to a future time when I have the privilege of examining another such book on this subject by an intrepid, inspired author who capitalizes even more on reviewing such items in explicit photographic detail.

AUTHOR'S NOTE ON PHOTOGRAPHY

Credits appearing below photographs are at times different but pertain to the same facility, abbreviated due to space constraints as the pages were laid out. For example, the Smithsonian Annex at Dulles Airport is properly called the Steven F. Udvar-Hazy Center. At times this has been shortened to Hazy Center or Smithsonian Annex. The Smithsonian National Air & Space Museum on the National Mall is also noted at times as Air & Space Museum or NASM. The National Museum of the United States Air Force in Dayton, Ohio, is sometimes USAF Museum or Dayton. Kennedy Space Center is abbreviated at times to KSC.

In many situations, pictures were taken through glass cases or clear Plexiglas panels in museum galleries, intended to show visitors hardware while avoiding any physical contact. Taking such photos requires some thought and planning. Taking them at an angle with flash can usually provide good results. Without flash, shielding the lens against reflections or putting the lens in contact with the panel can also be successful. Time exposures are fairly easy with a tripod, but at some facilities, tripods are not allowed, so using high ISO values, leaning against walls, or using handrails to balance the camera are ways to compensate. Action picture-taking such as rocket launches in daylight are a matter of following the object with a steady hand (a tripod is a big help here too) as it ascends and climbs from the launchpad. Due to safety and security concerns, you cannot get closer than a few miles away, so a zoom or telephoto lens makes a noticeable difference when viewing the end results.

For those dramatic night launch images showing the arc of the rocket path as it rises, this is probably the most challenging scenario to capture since you have only one chance to get it right. You must have the trusty tripod first and foremost. Aim the camera where you want the start of the rocket ascent to be, such as at the lower left corner of the viewfinder. Focus on something far off (or set the lens focus to infinity), then set the autofocus option to off or manual and do not touch the focusing ring again until the exposure is done. The f/stop setting is judged best by remembering that you are shooting into a black sky (in most circumstances) while a bright, intense flame trail crosses it. My best results have come at f/16 or f/18 at an ISO setting of 100 to obtain the best overall resolution possible. At liftoff, open the lens and let it expose until you feel you have enough of the action, then close the shutter to view the result.

Some cameras have a setting past the B (for "bulb") option that allows you to gently touch the shutter button to expose the digital sensor, then touching it again at your discretion to close it. Many cameras go down to only about thirty seconds for an automatic time exposure. This may not be long enough to catch all of what is happening. Even though you may have a wonderful arc of flame across the sky, you may miss the staging event, or solid boosters falling away. That is actually possible to capture if you leave the shutter open long enough. The best thing to do is to go out to a launch, try out some methods, and see which ones work for you. The pictures seen in this book are the best ones taken over many years of trial and error, but they are ultimately worth it once a collection of good shots and suitable methods begins to grow.

Chapter 1

THE BEGINNING: PROJECT MERCURY

The first artificial satellites launched were the Soviet Sputnik in 1957, and the American Explorer in 1958. The first American purpose-built spacecraft of the Mercury Project were specifically designed to carry men into space for extended periods. Shown in this chapter are the rockets used, launch facilities, early memorabilia, and preserved historic sites. A detailed examination of the Mercury spacecraft and manned launches is presented, and Liberty Bell 7, lost in 1961 and recovered in 1999, is also shown.

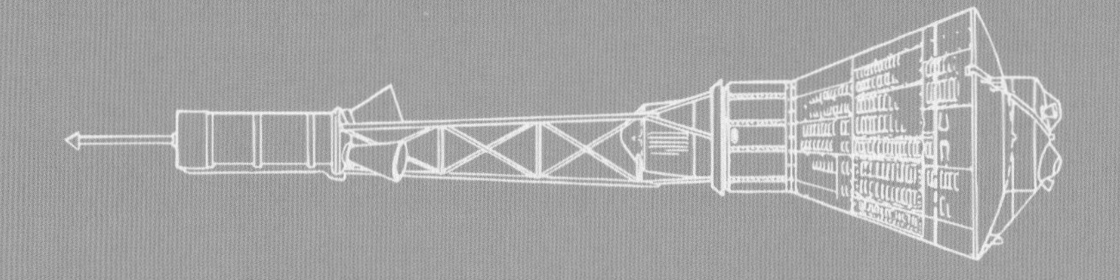

Left: a replica of Sputnik, the first man-made artificial satellite to orbit the earth, at the National Museum of the USAF in Dayton, Ohio. At right is the displayed replica of the American Explorer 1 satellite, the second man-made space traveler atop the Jupiter-C rocket, in a gallery surrounded by other rockets at the National Air & Space Museum in Washington, DC. With the launch of these, the space race was on, not only to send aloft larger, more-advanced satellites but to orbit a human in space as well. *Both photos by the author*

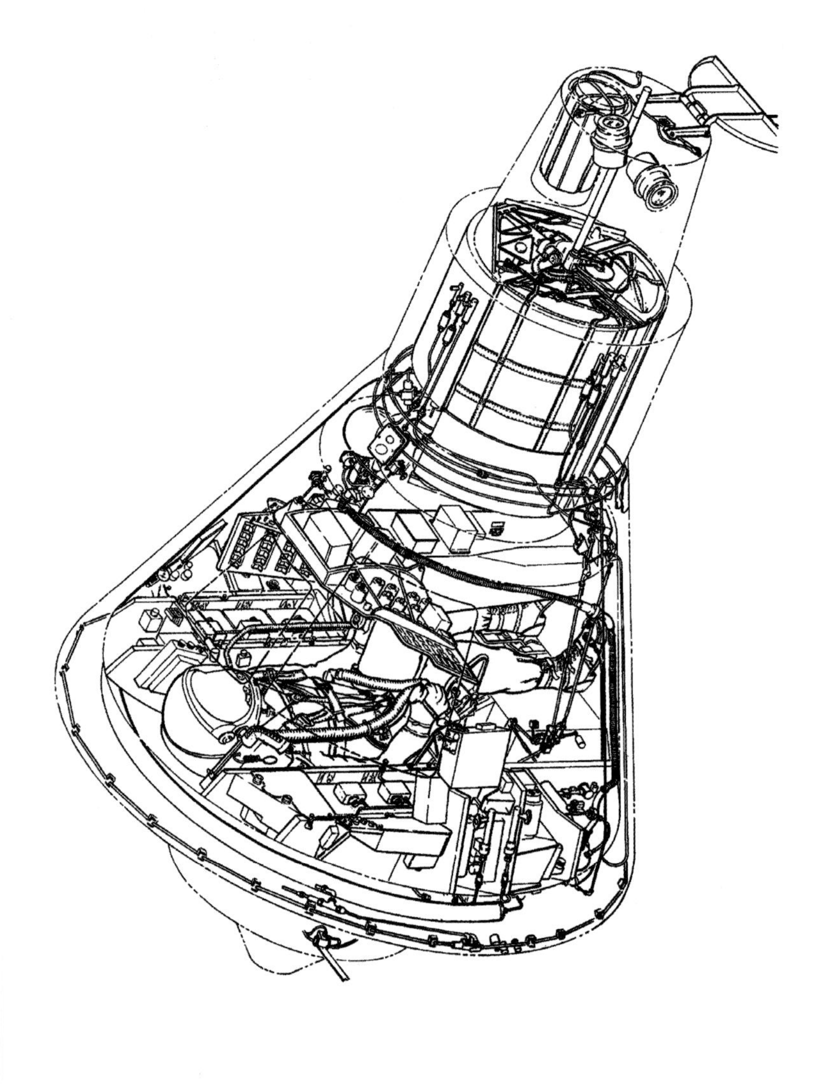

The left picture shows the Mercury Big Joe capsule that was flown on the first Atlas rocket flight test on September 9, 1959. The capsule was extensively instrumented for the thirteen-minute suborbital event and weighed 2,555 pounds, the largest payload yet flown. The test validated the heat shield and aerodynamic designs of the Mercury spacecraft, as well as postsplashdown recovery operations. The right line drawing is a page from the 1962 Mercury Familiarization Manual, showing the complete manned layout, with an astronaut in his contoured couch, instrument panels, and flight hardware as the spacecraft would appear in orbit. *Left photo, Steven F. Udvar-Hazy Center by author/right illustration, NASA*

Mercury Procedures Trainer at the Huntsville, Alabama, Space & Rocket Center. Used by the Mercury 7 astronauts for familiarization with the interior layout and interacting with the various systems on board.

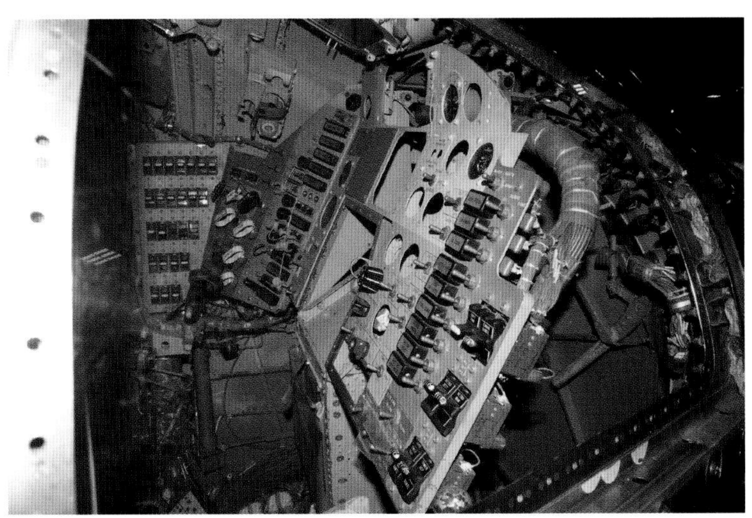

Interior view of the Mercury Atlas-10 spacecraft Freedom VII II, which did not fly. Panels are colored according to groups of dials, switches, and indicators. Green panel in foreground could be removed if the astronaut had to egress through the forward section.

Mercury capsule at the National Museum of the USAF was intended for flight but instead became a source for spare parts. The heat shield has been removed, showing a toroid-shaped hydrogen peroxide tank.

Forward section of the USAF Museum Mercury shows empty bays where recovery parachutes were packed. Corrugated bulkhead in background was the removable emergency escape hatch. Astronaut passed through empty bay section to egress if necessary. *Photos this page by author*

Liberty Bell 7 was lost on July 21, 1961, after the second suborbital flight. It was eventually recovered on July 20, 1999, after thirty-eight years of saltwater immersion. It is seen while on a tour stop at Kennedy Space Center in July 2000.

Liberty Bell 7 interior view. The spacecraft has been meticulously restored by the staff of the Kansas Cosmosphere and Space Center. Large round display was used when the periscope was deployed. Various pull handles, annunciator lights, and gauges are also seen.

When a Mercury capsule landed in the ocean, the heat shield lowered to deploy a landing bag. Upon splashdown, air in the bag exited through the large holes, providing a shock absorber effect to cushion the impact.

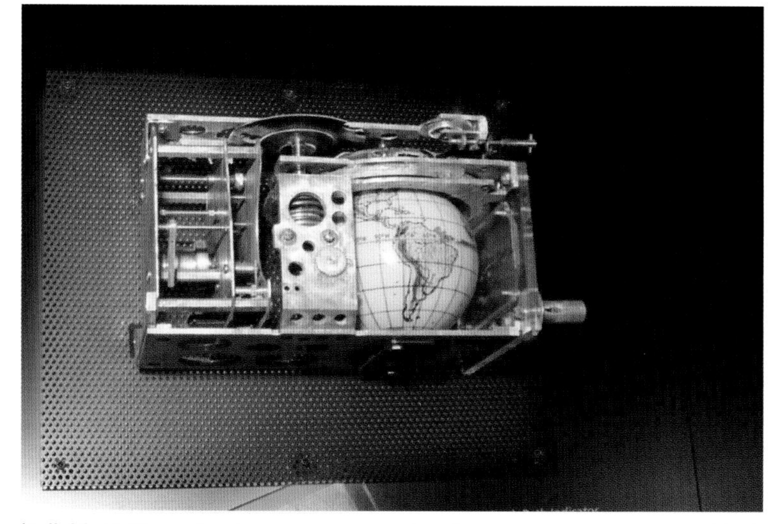

Individual instruments were removed, taken apart, and reassembled. This is the Earth Path Indicator that Gus Grissom referred to as his flight progressed. As the spacecraft moved, the globe would move to provide situational awareness. *Photos this page by the author*

The spot from Launch Complex 5 where Alan B. Shepard lifted off on his suborbital flight. This was the first successful launch of an American astronaut, and it took place where this Mercury-Redstone display stands at Cape Canaveral Air Force Station in July 2009. After the two suborbital proving flights of Shepard and Grissom, a larger rocket was necessary to achieve the capability of orbiting the earth, so the Atlas intercontinental ballistic missile (ICBM) was adapted or "man-rated" for the role. This example is part of the Kennedy Space Center rocket park in June 2013. Size comparison with the Redstone in the background is noteworthy. *Both photos by the author*

The blockhouse at Complex 26 is adjacent to Launch Complex 5 and is constructed of poured concrete, with thick windows to observe the launchpad action. The roof has a dish antenna for telemetry reception.

A definitive blockhouse design on Cape Canaveral Air Force Station. Periscopes were used to watch pad events. The design was so strong that even if a rocket impacted the structure, the occupants would be unharmed.

The interior of the Complex 26 blockhouse, with the mission director console in foreground. Racks full of computers, radios, and technician and engineer consoles fill the limited space with a view of the launchpad.

At Complex 31, another blockhouse design involved layering the exteriors with sandbags, giving the appearance of a beehive, which is what they remain nicknamed today. Access to the top as well as the entrance at left is seen here. *All photos this page by the author*

John Glenn's capsule, Friendship 7, on display at the Smithsonian National Air & Space Museum. It is encased inside Plexiglas for public viewing. It completed three orbits of the earth on February 20, 1962.

This button was pushed to send John Glenn into orbit on February 20, 1962. Seen at the American Space Museum in Titusville, Florida, in January 2019. A small but significant item of US space achievement.

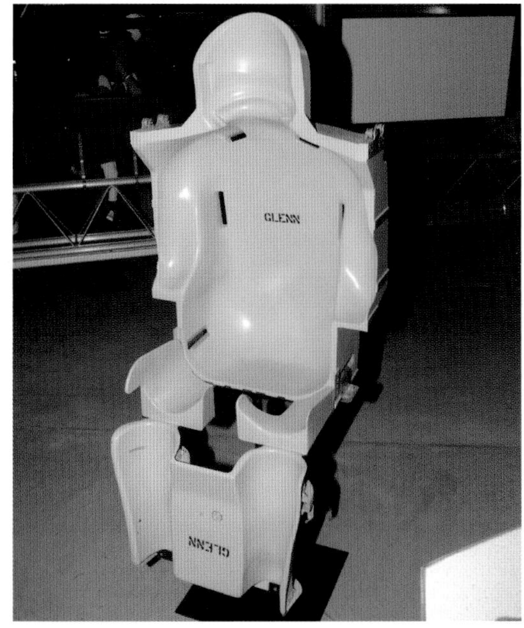

Each of the Mercury 7 astronauts had a contoured couch custom-fit for them. This mold was used to create John Glenn's couch for his historic flight. Seen at the Stephen F. Udvar-Hazy Center in July 2013.

After Glenn's flight, many items of memorabilia were produced, such as the commemorative plates, postcards, patches, and models seen at the American Space Museum. *All photos this page by the author*

The Smithsonian Annex at Dulles International Airport (also known as the Steven F. Udvar-Hazy Center) is where one example of a complete Mercury capsule in orbital configuration is displayed. It is the Freedom VII II spacecraft that Alan Shepard would have piloted had the mission been authorized to fly. It was intended to be of extended duration in earth orbit, but the success of the Mercury-Atlas 9 flight of Gordon Cooper (Faith VII of twenty-two orbits in May 1963) resulted in future efforts devoted to the two-man Gemini Project. Photo shows the modified retrorocket package secured to the heat shield, providing extended life-support capabilities. This pack was jettisoned before reentering the atmosphere. The front fairing with red cover contained the drogue parachute, and a pair of horizon scanners (one for pitch and one for roll) were used to maintain correct capsule attitude while orbiting. Normal entry hatch is removed, allowing visitors to view the interior. *Photo by the author*

MERCURY PROGRAM MANNED FLIGHTS

Date	Mission	Astronaut/Revolutions	Duration
May 5, 1961	Mercury Redstone 3 / Freedom 7	Alan Shepard / suborbital	15 minutes
July 21, 1961	Mercury-Redstone 4 / Liberty Bell 7	Gus Grissom / suborbital	15 minutes
February 20, 1962	Mercury-Atlas 6 / Friendship 7	John Glenn / 3 orbits	4 hours, 55 minutes
May 24, 1962	Mercury-Atlas 7 / Aurora 7	Scott Carpenter / 3 orbits	4 hours, 56 minutes
October 3, 1962	Mercury-Atlas 8 / Sigma 7	Wally Schirra / 6 orbits	9 hours, 13 minutes
May 15, 1963	Mercury-Atlas 9 / Faith 7	Gordon Cooper / 22 orbits	1 day, 10 hours, 20 minutes

Mercury program total cost: $277,000,000 (in 2019 dollars, the total cost would have been $2.25 billion)

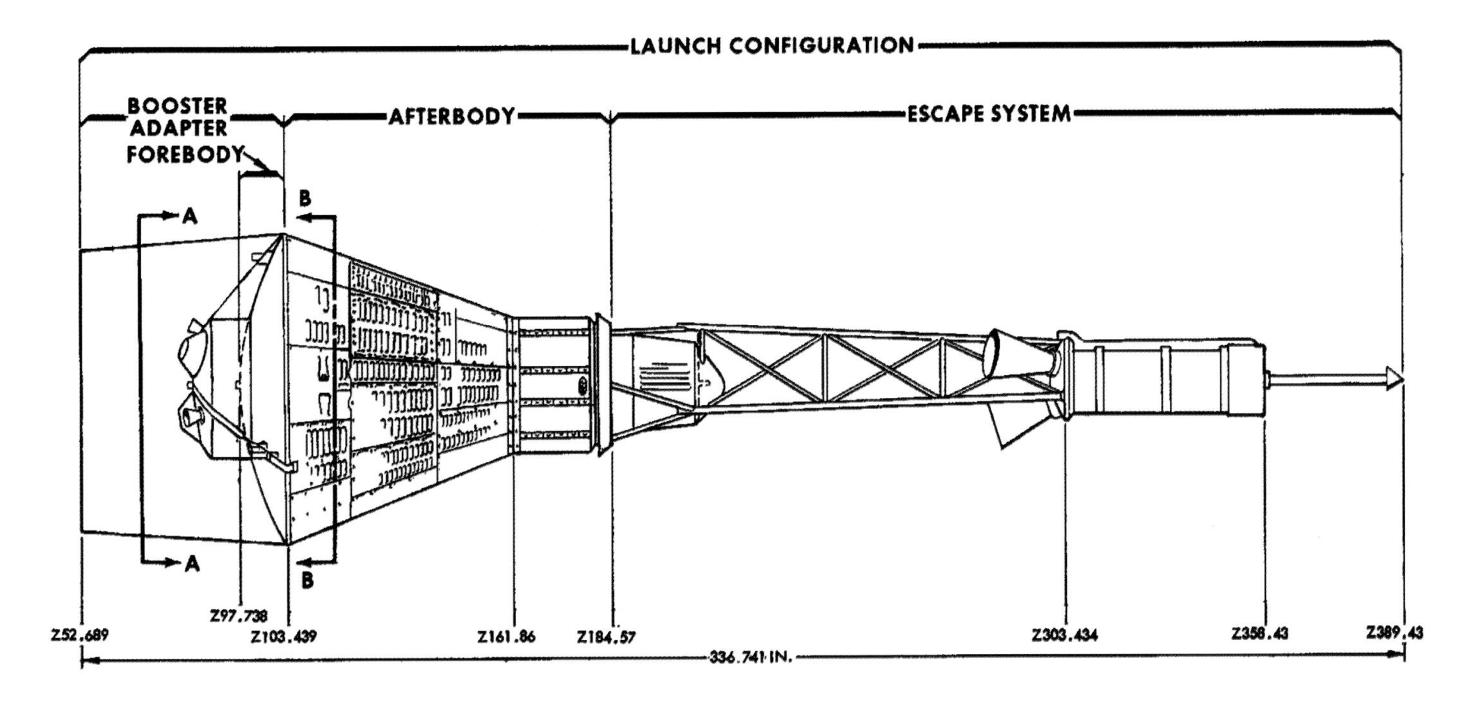

The Freedom VII II forward fairing shows round cover for the drogue parachute, and the center tube is for the fairing ejection gun. The horizon scanner under red cover maintains pitch orientation; the forward scanner above in image is used for roll alignment. *Hazy Center*, *photo by author*

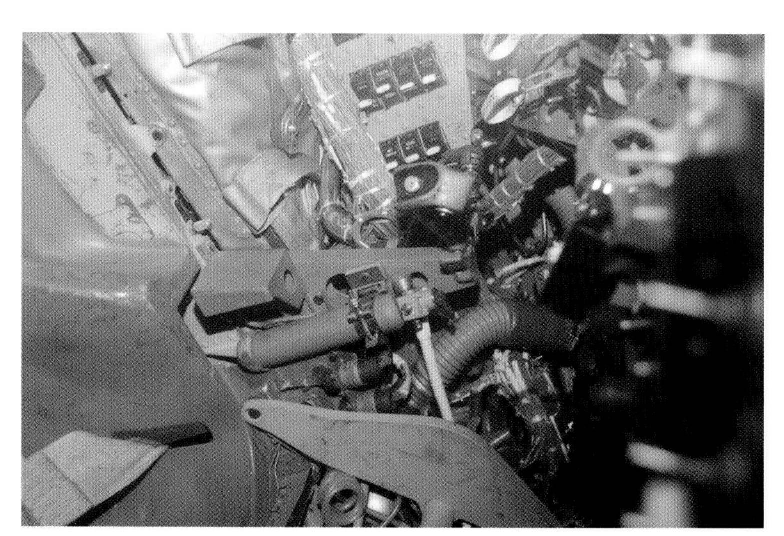

Sigma 7 interior shows the couch left-hand abort handle. The red button is pushed and the handle is moved outboard to initiate the abort sequence. The green cylinder supplies oxygen to the face lens seal of the helmet. *Astronaut Hall of Fame*, *photo by author*

Faith 7 heat shield displayed at the Astronaut Hall of Fame in Florida. The holes are where samples of the ablative material were removed for further study. The Faith 7 flight was the one in which twenty-two earth orbits was achieved. *Author*

Sigma 7 three-axis hand controller on right-hand side of couch. The handle was twisted to input yaw corrections. Moving it forward/aft made changes in pitch. Side-to-side motion would change orientation in the roll angle. *Author*

Complex 14 at Cape Canaveral Air Force Station, where the four Mercury-Atlas launches occurred. The area has many placards noting historic events at the various launchpads seen on tours of the facility. *Author*

The Mercury program's time capsule, located 2,200 feet west of the Complex 14 launchpad, to be opened in the year 2464, five hundred years after it was dedicated in 1964. Technical reports of the missions flown are contained within. *Author*

The Mercury Program Memorial at Cape Canaveral Air Force Station. The astronomical symbol for the planet Mercury has the numeral 7 in the center, to commemorate the seven astronauts who were involved. *Author*

At the entrance to Complex 14 are ceremonial parking spaces for the original seven astronauts. Seen here are those for Scott Carpenter, Wally Schirra, and Gordon Cooper. Today, the cape is part of the United States Air Force Space Command. *Author*

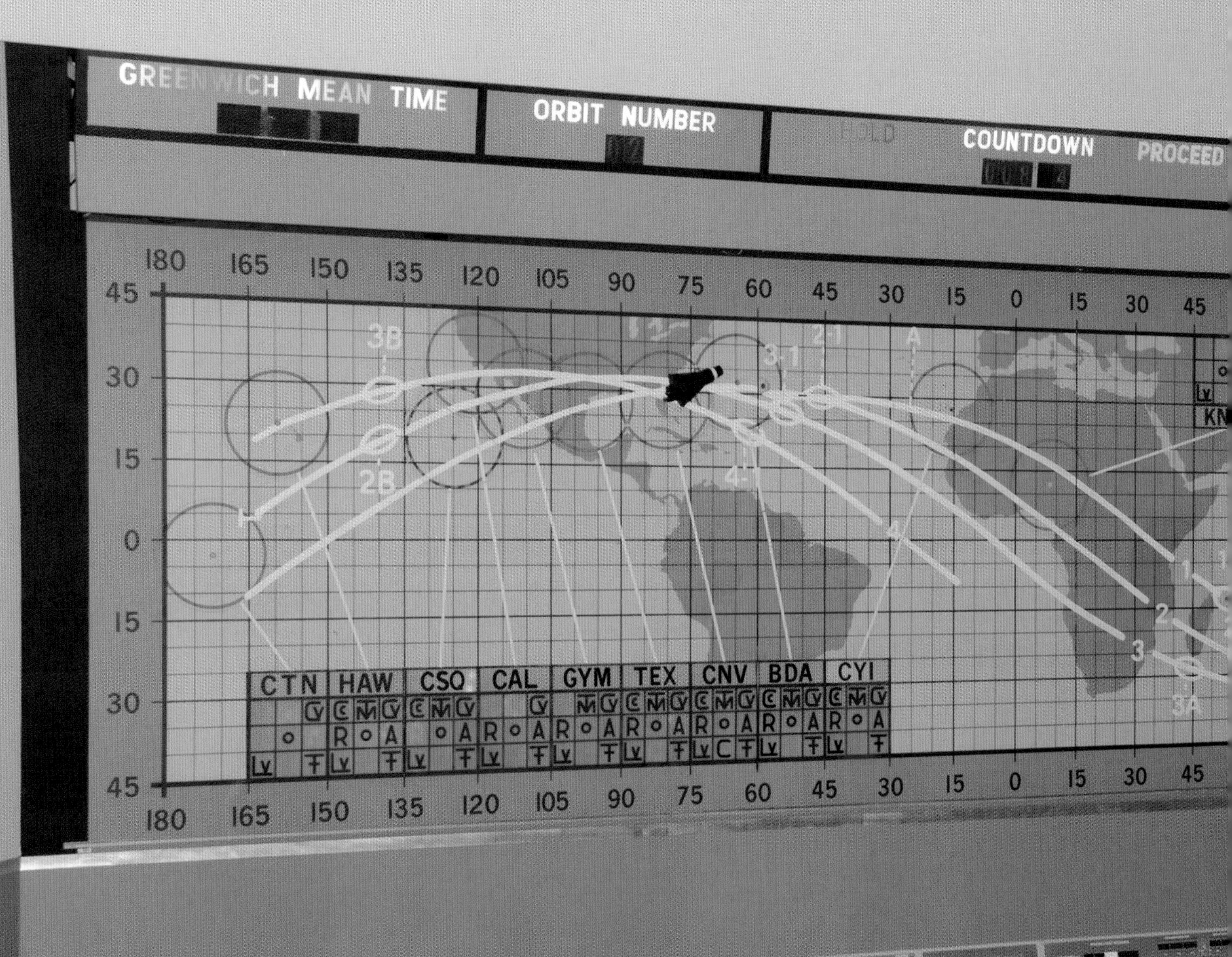
GREENWICH MEAN TIME
ORBIT NUMBER
COUNTDOWN
PROCEED
180 165 150 135 120 105 90 75 60 45 30 15 0 15 30 45
CTN HAW CSQ CAL GYM TEX CNV BDA CYI

SANBORN RECORDER NO. 2

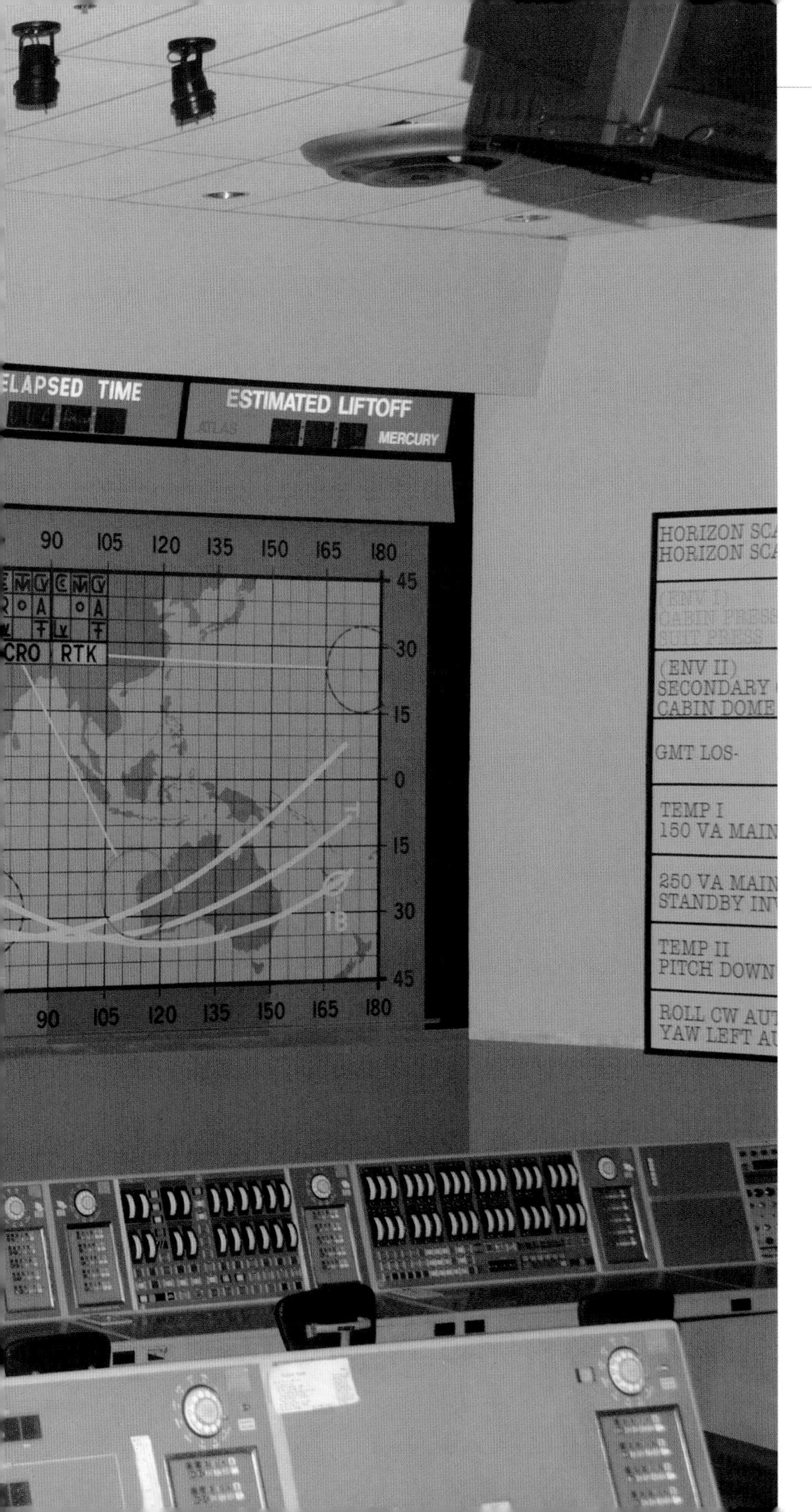

Project Mercury Mission Control Center is preserved at the Kennedy Space Center, Florida. Engineers and technicians manned the consoles during flights, monitoring data unique to their field, such as life support, capsule systems such as status of maneuvering propellant, and communications links. The world map showed where the spacecraft was (black capsule shape moved across the board as it orbited the planet), with tracking/telemetry stations surrounded by black circles, showing their areas of optimal communications and telemetry reception. White lines to charts identified which stations were available, such as HAW for Hawaii, CNV for the Cape, and BDA for Bermuda. Thick white lines showed orbits accomplished and the next one to be made. *Photograph taken in July 2013 by the author*

Chapter 2

THE GEMINI PROJECT

With the conclusion of the Mercury Project, attention focused on the Gemini series of missions. The first test flight occurred on April 8, 1964, from Launch Complex 19 at Cape Canaveral, Florida; a capsule was placed in orbit and was not recovered. A second launch on January 19, 1965, resulted in the recovery of the Gemini capsule, placed on a suborbital trajectory to test the heat shield design. Post-recovery, it was overhauled and made flight-ready for another launch, designated OPS-0855/OV4-3. That capsule, Gemini 2, has been preserved at Cape Canaveral Air Force Station and was the only reused spacecraft, other than the X-15, for many years until the arrival of the reusable space shuttle orbiters. Another modification that did not make it into use was for a Parawing (an inflatable design invented by Francis and Gertrude Rogallo, known also as a flexible or Rogallo wing), which was to be used for controlled ground landings instead of parachutes, with a Gemini capsule equipped with wheeled landing gear. This was actually tested but ultimately abandoned as an option, relying instead on the proven ocean splashdown technique. This configuration is displayed at the Smithsonian Annex / Udvar-Hazy Center.

A modified Gemini capsule of the cancelled MOL Project is also on public display at the National Museum of the United States Air Force, in Dayton, Ohio. Looking at the unflown Ohio example, known as Gemini B, it looks like the rest of the displayed Gemini capsules, but on closer examination it is seen that a heat shield hatch allowed for the crew to open this and enter the MOL platform for extended missions of up to thirty days. The capsule could remain docked for as long as forty days of space exposure. The ejection seats were angled slightly farther outboard to compensate for the hatch installation. Internal systems were modified to meet conditions associated with periods of minimal use while docked to the MOL. The heat shield was made more robust to cope with the increased heat levels and atmospheric loads placed on it during reentry profiles from polar orbits. The first manned launch, on March 23, 1965, was Gemini 3, a three-orbit test flight of four hours, fifty-three minutes, with a crew of astronauts Virgil I. "Gus" Grissom and John Young. The capsule had a computer installed that could be programmed by a crewman to make orbital adjustments while circling the Earth. Grissom did so and was able to perform the first-ever orbital maneuvers.

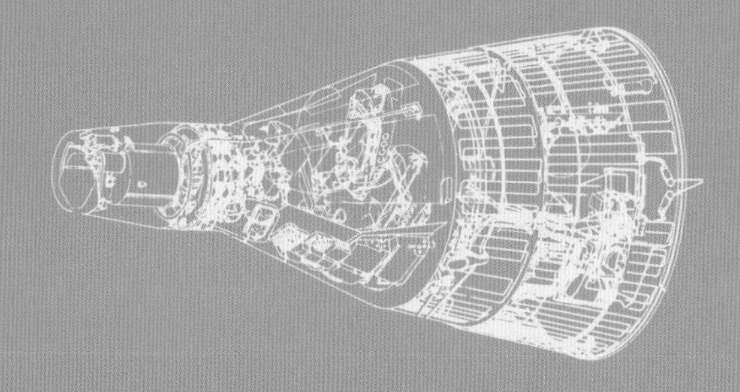

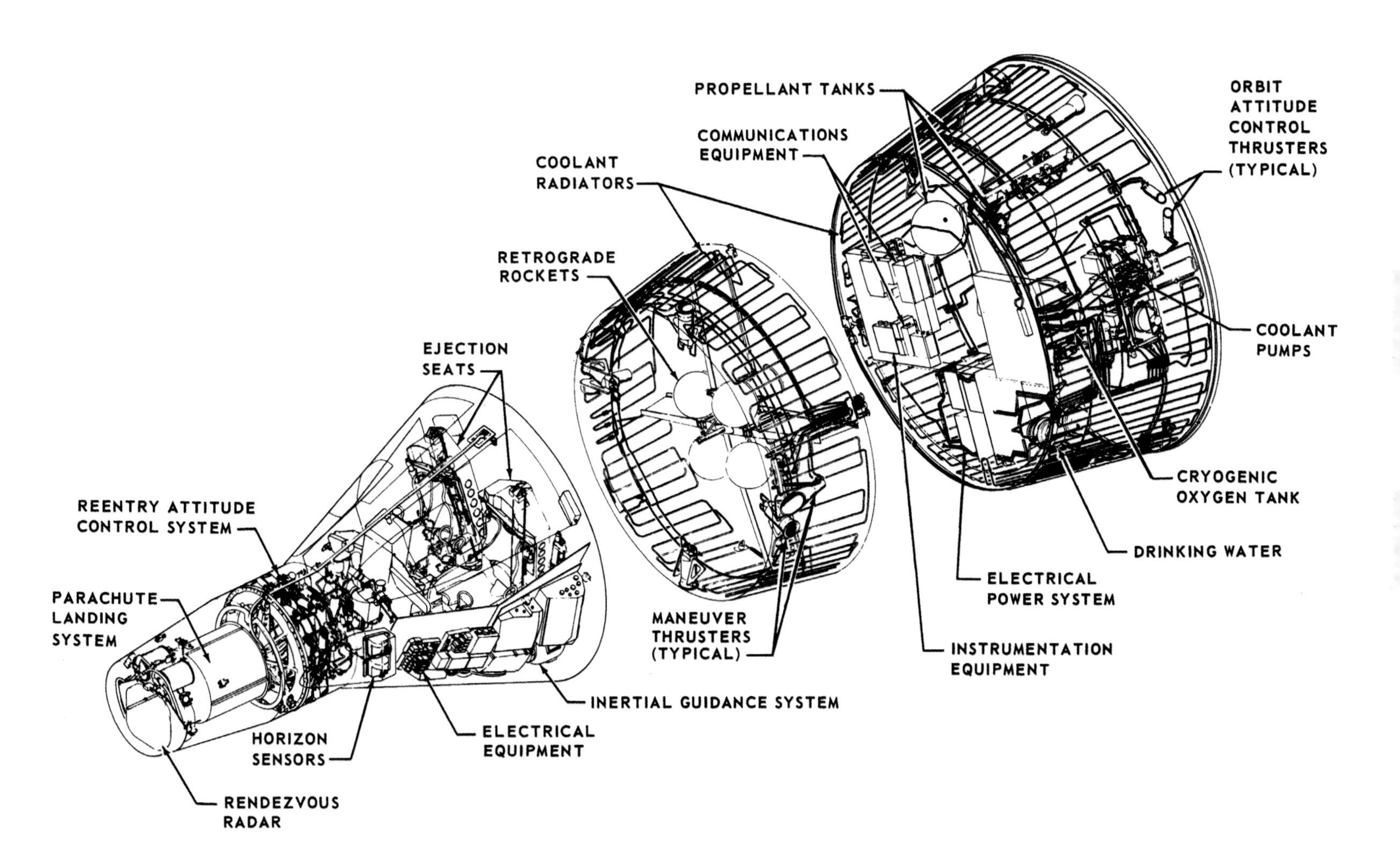

Gemini spacecraft cutaway diagram, showing major assemblies of the rendezvous and recovery nose section, the reentry attitude control system ahead of the two-man crew cabin. Retrograde section was attached aft of the heat shield and the adapter section containing all the supporting equipment required for long-duration stays in space. *NASA*

Gemini 2 at Cape Canaveral Air Force Station in June 2013. Still visible is the "US Air Force" and star-and-bar roundel, as opposed to "United States," on all the NASA capsules.

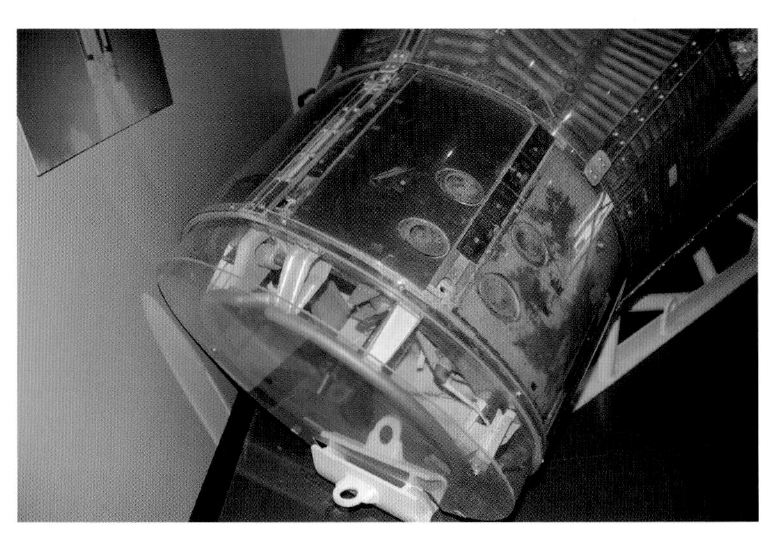

Gemini 2 reentry section containing a total of sixteen 25-pound attitude control thrusters. This is the only assembly that came back with the capsule cabin section at the time of splashdown.

Gemini 2 left-hand seat section of cabin had, in place of the ejection seat, bars of ballast to simulate the two crewmen, sensors, relays, solenoids, telemetry, and recording gear.

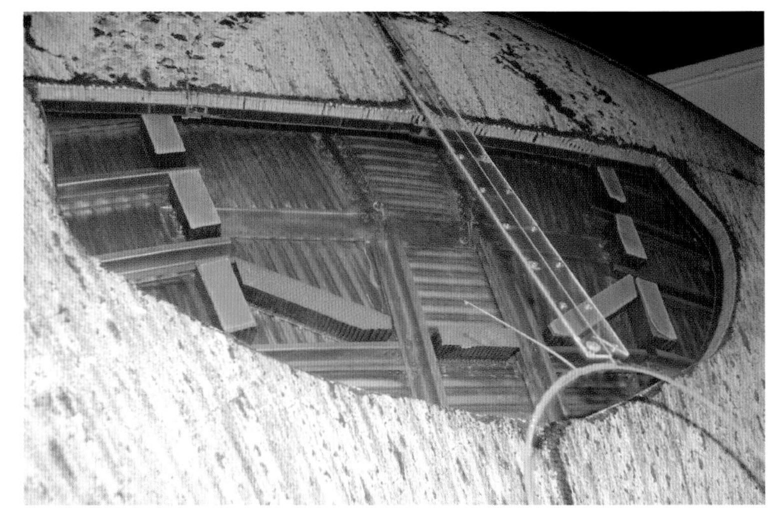

Since this capsule was used twice in space, the heat shield was of particular interest to engineers. A large section has been cut away, revealing the supporting structure and spacers on which the shield rested. *All photos this page by author*

Gemini B, intended for the Manned Orbiting Laboratory (MOL) Project. This capsule never flew in space but had been modified as a MOL item. Entire capsule is encased in clear Plexiglas at the US Air Force Museum in Dayton, Ohio.

The thicker heat shield also had the access hatch that the pair of astronauts would open to enter the MOL orbiting station. Stays in space were expected to last thirty days, after which they would occupy the capsule, leave the outpost, and land.

Interior view of Gemini B shows the hatch with six latch arms and operating handle. Crewmen would open hatch and enter the MOL, the task being made much less strenuous by floating through the opening in zero gravity.

View of the forward end of Gemini B shows a green item attached to the inner wall of the reentry section. This is a simulated green-dye marker that would, after immersion in water, turn the sea a green color visible to recovery crews. *All photos this page by author*

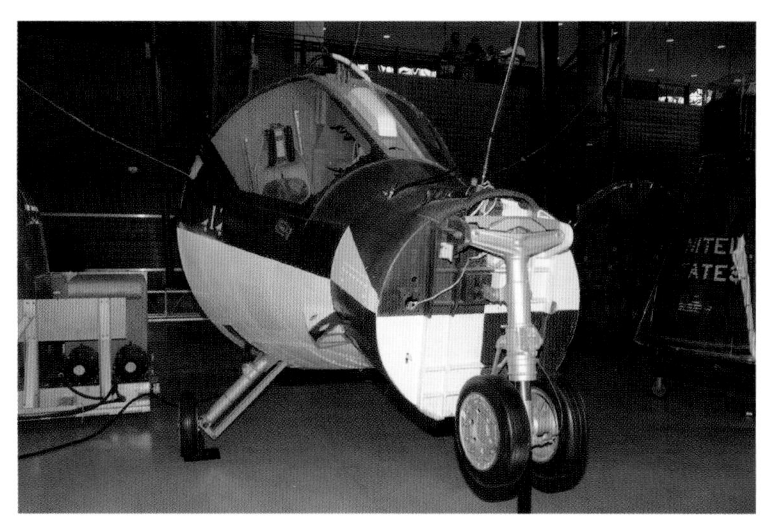

The Test Tow Vehicle (TTV-1) was built to resemble a Gemini capsule, with ejection seats, outfitted with wheeled landing gear. Black paint reduced glare into the eyes of the crew. Red and white paint increased visibility.

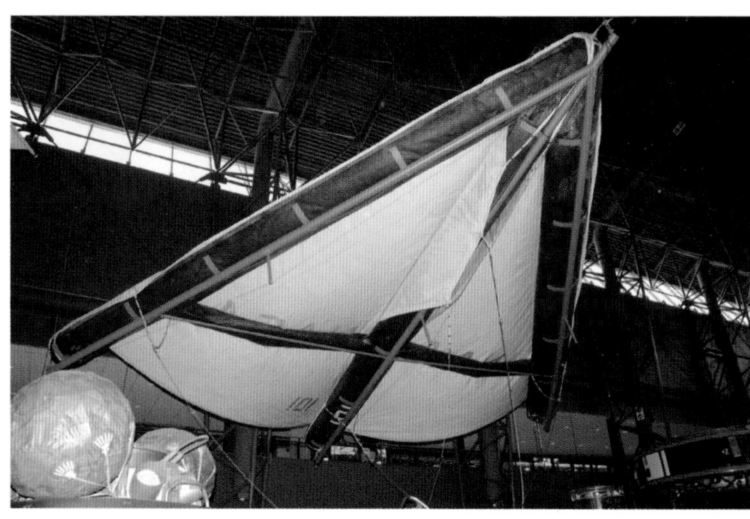

The Rogallo wing above TTV-1 at the Steven F. Udvar-Hazy Center. It would have been stowed inside the capsule, inflated after reentry, and used to glide in a controlled state to an eventual landing on the ground.

TTV-1, reverse angle. Visible are three white mission markings painted on the side. Vehicle was taken to altitude by helicopter and released, with the wing already inflated and deployed.

This view from behind TTV-1 shows the high-visibility paint design on the heat shield, the wing planform, and the size of the wing itself, in relation to the capsule. It was a promising concept but complex in practice. *All photos this page by author*

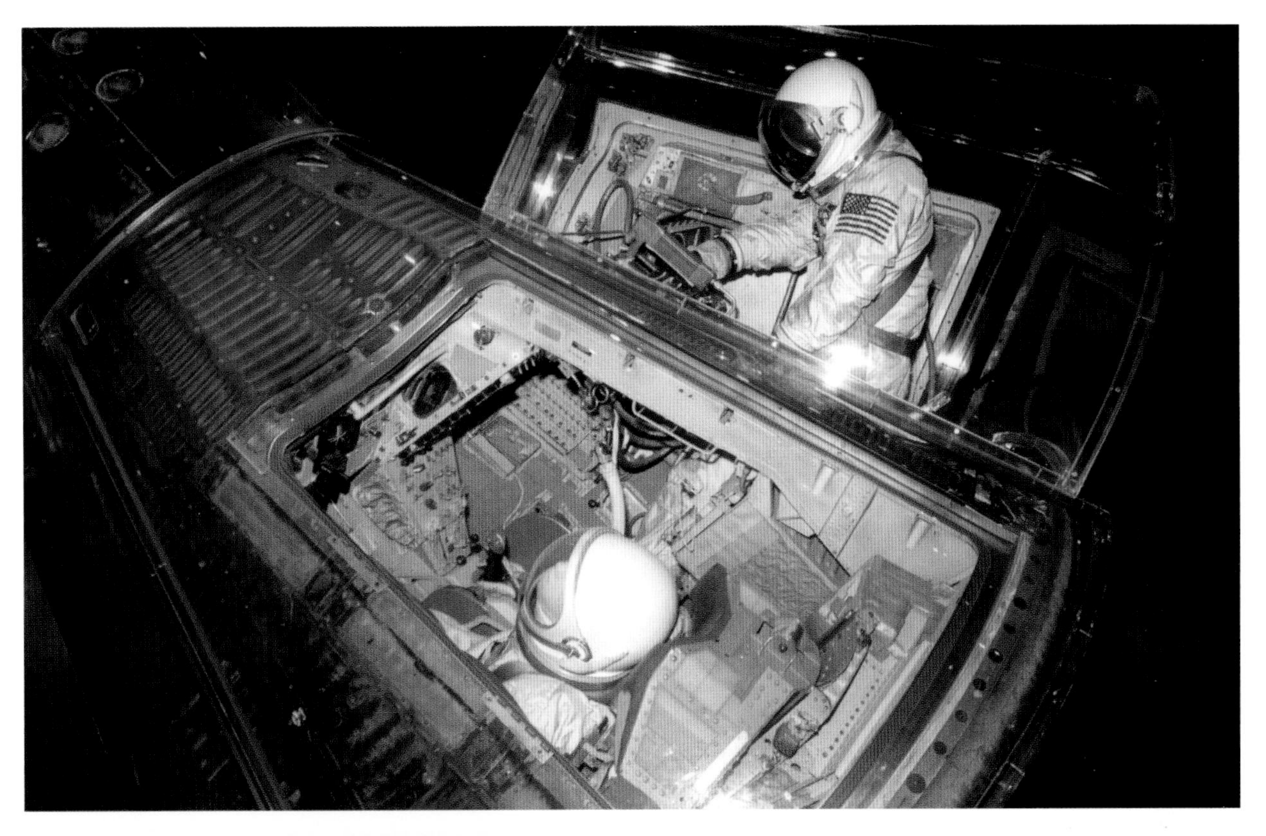

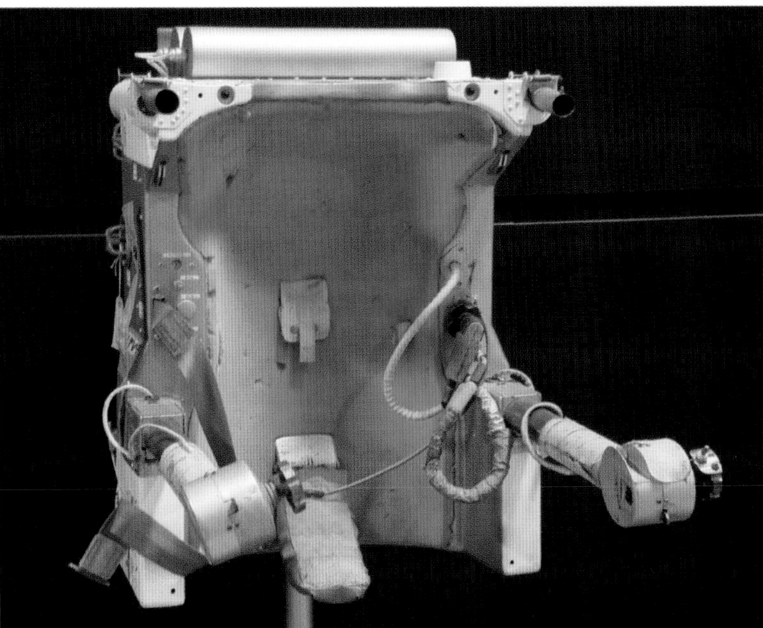

The Gemini 4 capsule, displayed at the National Air & Space Museum in December 2002, depicts the scene during the third earth orbit; in June 1965, Ed White floated out of the cabin to become the first American to perform a spacewalk, Russian cosmonaut Alexei Leonov having done so in March. The hatch over James McDivitt's seat has been removed. A mannequin representing White has in the right hand the maneuvering unit that gave him some control as he spent twenty minutes outside the spacecraft. Oxygen and communications were supplied through the gold-foil-wrapped umbilical cord, also visible. *Author*

The flight of Gemini 9 carried an Astronaut Maneuvering Unit (AMU) at the back of the spacecraft. It weighed 165 pounds (75 kg). It was 32 inches (81 cm) high, 22.5 inches (56 cm) wide, and 19 inches (48 cm) deep. A tether of 147 feet (45 meters) was attached from the unit to the spacecraft as a safety measure while testing the unit prior to flight. Twelve thrusters provided maneuvering capability, using hydrogen peroxide gas. The pair of arm supports were also the means to control the AMU in space. It also contained a telemetry link, communications, life support, and an automatic stabilization system. *Air Force Museum, Dayton, Ohio, October 2018, by the author*

Seen at the Space & Rocket Center in Huntsville, Alabama, are Gemini-era boots and helmet. The silver pair go over the boot in the background to reflect the sun's radiation, keeping the astronaut cool. *Photos this page by author*

This camera would have filmed Ed White's spacewalk if the identical camera were unsuited for flight. It was attached to the spacecraft, using a wide-angle lens to view White, the capsule, and the earth. *Astronaut Hall of Fame*, *March 1994*

Ed White was able to do some maneuvering with this handheld gas unit. In actuality, it ran out of propellant during the walk, so White moved about by pulling gently on the gold-wrapped umbilical cable. *Huntsville, October 2016*

A view inside the cabin of Gemini VII, looking at the command pilot position in the left seat. The suit displayed is one of the lightweight versions designed to provide comfort during extended flights. *Air & Space Museum*, *December 2002*

The left image shows a Gemini capsule atop a Titan II rocket, adapted from the ICBM production line. Man-rating it involved, among other items, engine reliability upgrades, batteries that could be recharged, and structural modifications. The right photo shows an Atlas rocket with an Agena docking vehicle as the payload. These were sent into orbit in order for astronauts to conduct approaches to, rendezvous with, and then attach to temporarily. While docked in orbit, procedures and tests could be conducted. A less sophisticated substitute docking vehicle, the ATDA, or augmented target-docking adapter, was also used as a substitute for the Agena. *Kennedy Space Center Rocket Park, July 2013, photo by author*

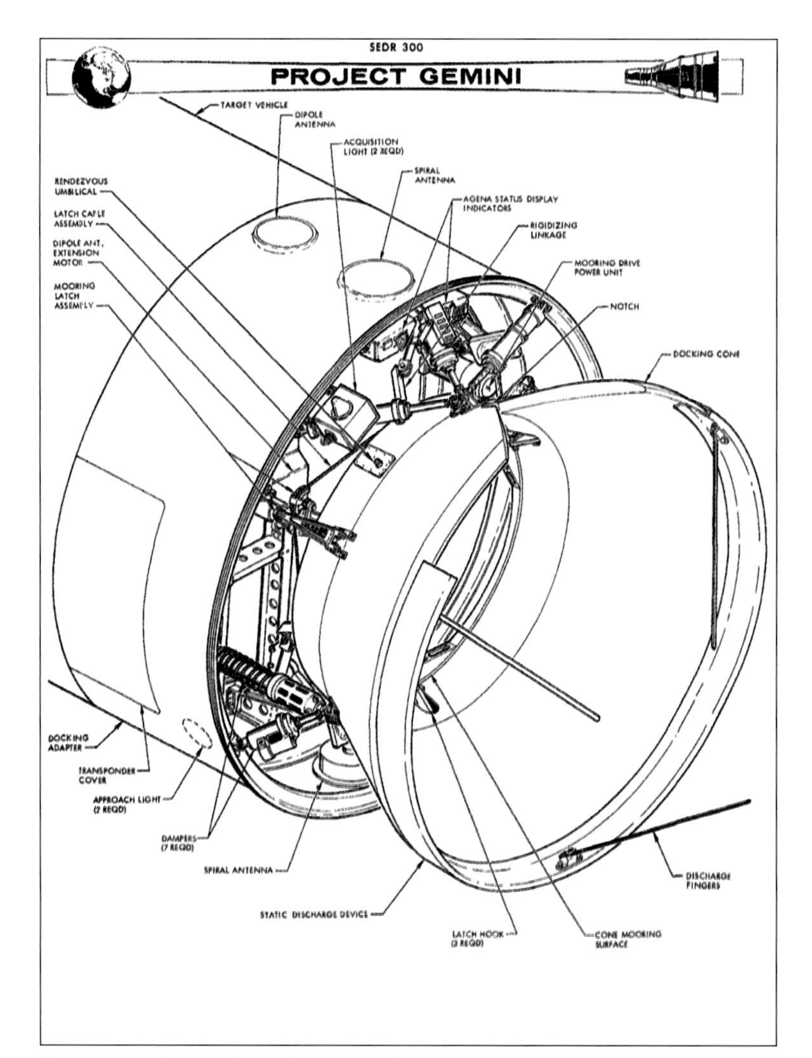

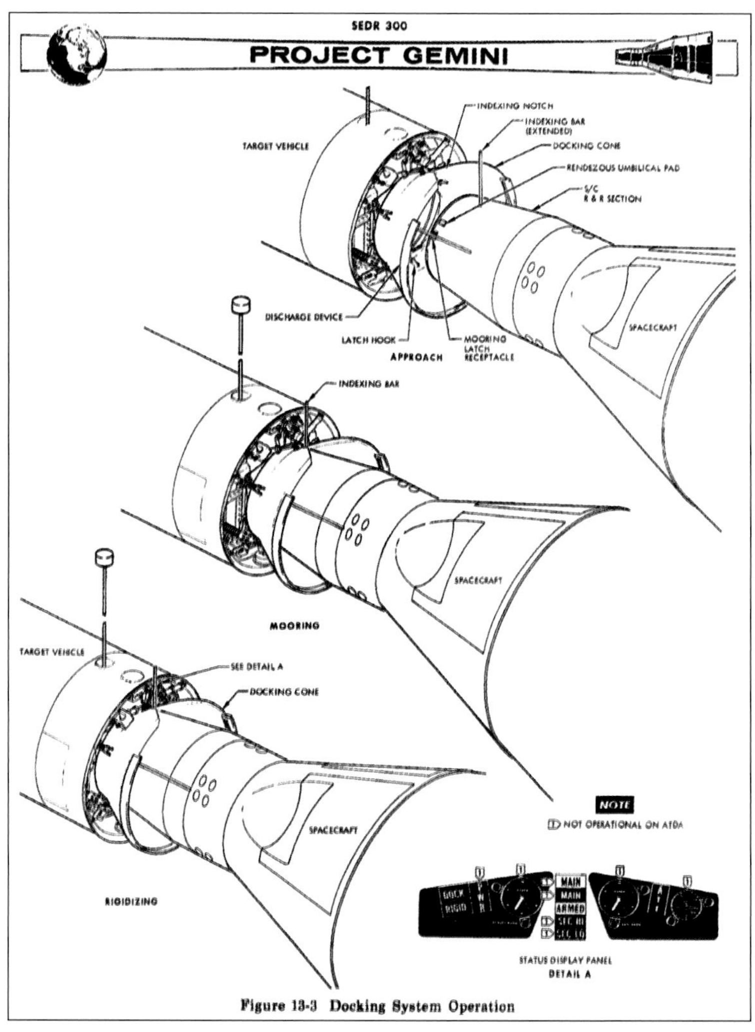

Figure 13-3 Docking System Operation

The Agena forward end had a docking cone that the capsule forward section fit into after a rendezvous maneuver was completed. Three discharge fingers touched the capsule initially to dissipate any static electricity buildup as docking began. Ahead of the notch where the indexing bar fit into the cone were panels of gauges and indicators the astronauts could view to evaluate docking status. The Agena had an engine that the crewmen could operate to change the orbit of the mated vehicles. Docking/undocking was a major goal of the Gemini Project, an essential task to master if any chance of attempting a moon landing was to be considered. *Both diagrams, NASA*

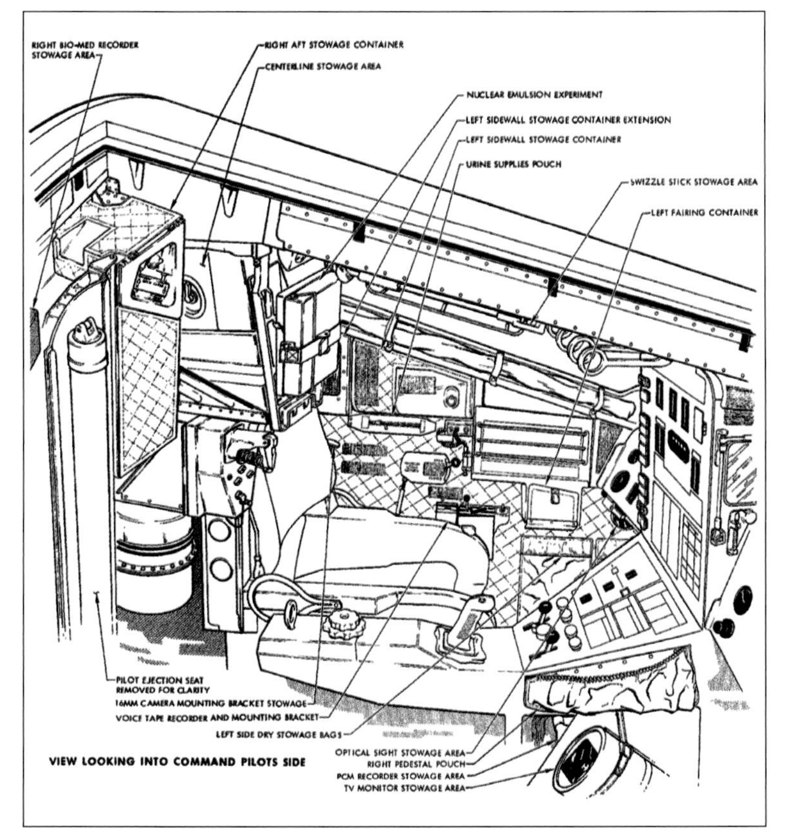

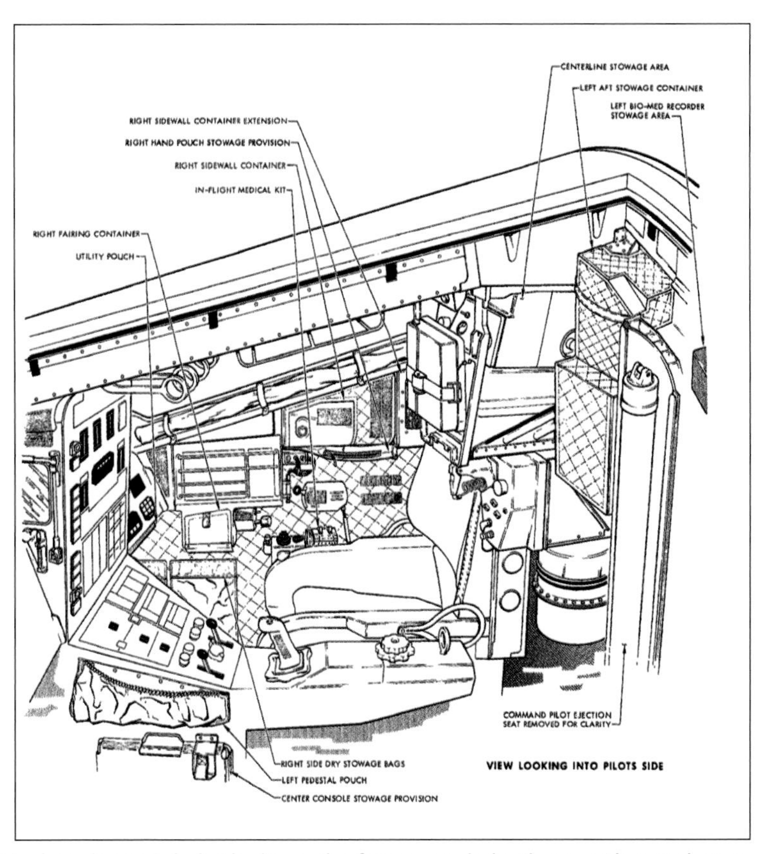

A pair of NASA line drawings of the Gemini interior crew cabin show highlighted equipment. In both views, the foreground ejection seat is not shown, to provide clarity into cabin sections. Loop-shaped handles folded aft on each side of seats are arm restraints designed to orient astronauts into proper body alignment prior to ejection. Central attitude control handle could be used by either crew member. Round knob aft of hand controller is for cabin temperature control. Upper-row lineup of five ring-shaped levers are for (*left to right, facing forward*) cabin air recirculation, snorkel, cabin vent, water seal, and oxygen high-rate recock. The upper assembly of the main instrument panel's central console has gauges for oxygen and propellant fluid levels, the UHF radio, telemetry, and beacon control, ECS (environmental control system) heaters. The lower assembly of the main instrument panel center console has interfaces for rendezvous radar control, computer power / mode select, attitude control, rate gyros, and OAMS (orbit attitude maneuvering system) and RCS (reaction control system) controls. The "swizzle stick" was used by one astronaut to operate switches on the other side of the cabin if the other occupant was asleep or busy performing a task.

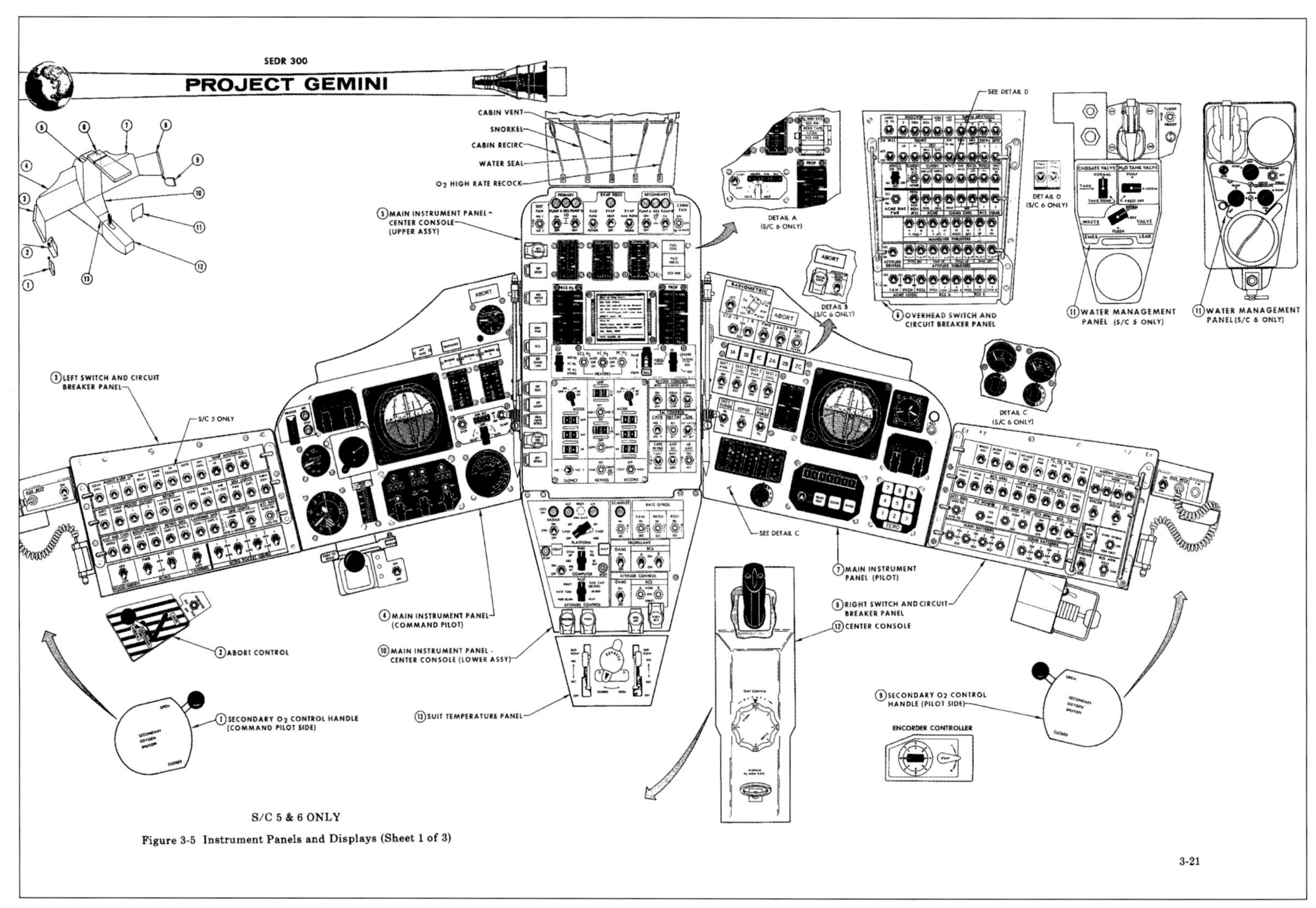

S/C 5 & 6 ONLY

Figure 3-5 Instrument Panels and Displays (Sheet 1 of 3)

This NASA layout shows Gemini V and VI instrument panels. There were slight differences in instrumentation throughout the program as technology improved or mission requirements dictated that other items be included. Abort control and secondary oxygen control handles are located on cabin sidewalls. The water management station was positioned on the aft cabin bulkhead, between the ejection seats, where both astronauts could easily access it. Differences in a crew member's main instrument panels are that, for example, the command pilot has an altimeter gauge and rate-of-descent meter. Both have attitude indicators that dominate their arrays of instruments.

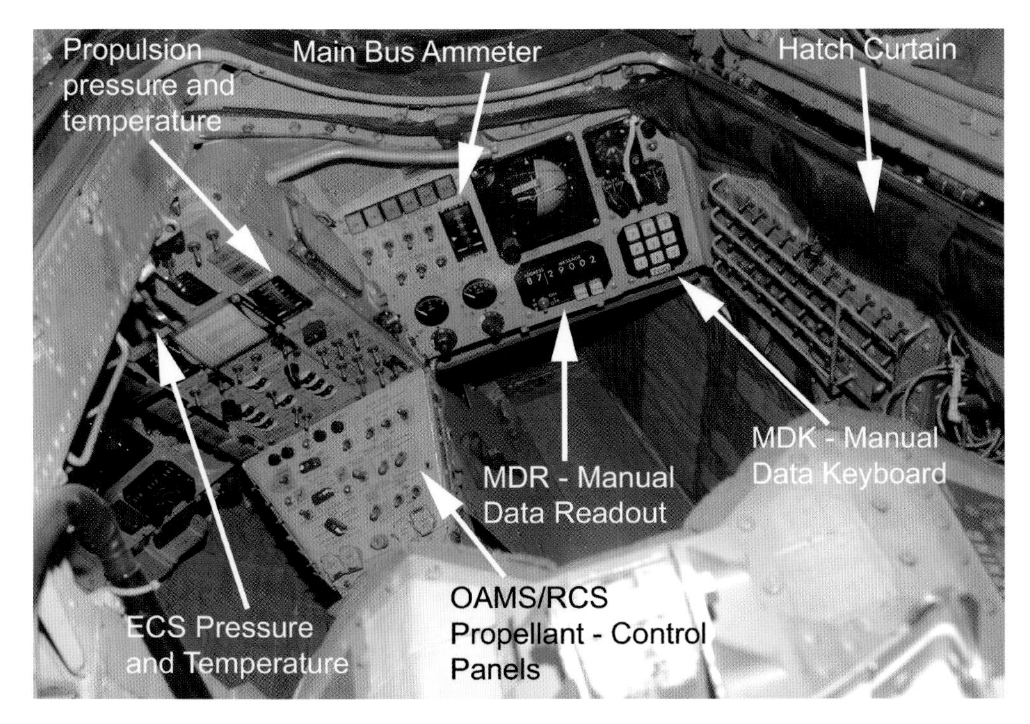

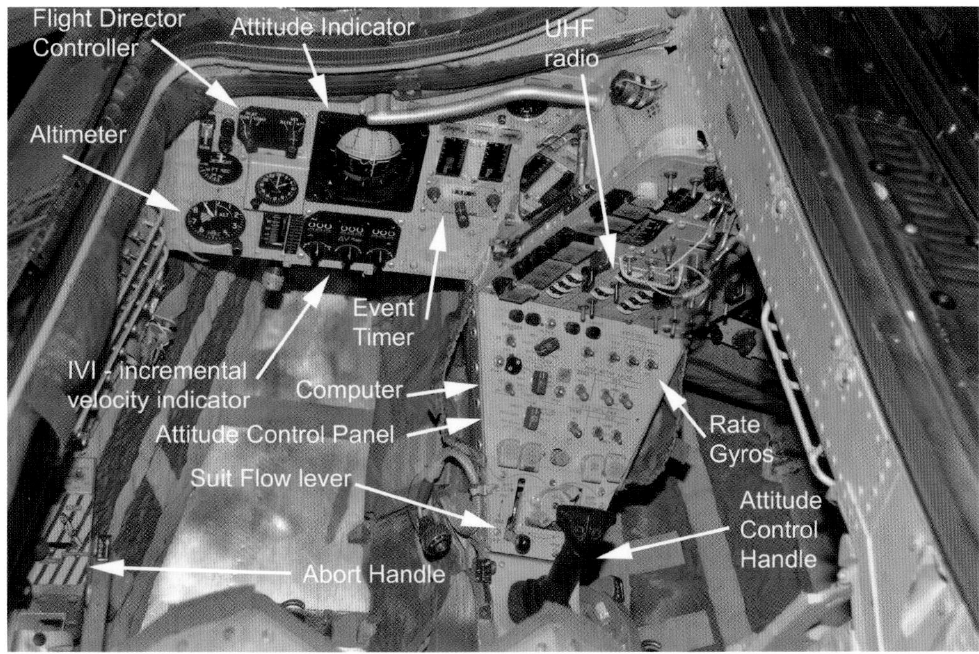

In the top image, taken at the Smithsonian Air & Space Museum of the Gemini IV pilot station, various items are highlighted. Each astronaut had identical instruments on his dedicated flight panel, with others unique to their stations. The hatch curtains were raised after splashdown to prevent seawater from entering the cabin, should excessive wave action be present. OAMS and RCS (orbit attitude maneuvering system, reaction control system) panels deal with the nose section / aft section thruster arrays and fuel status. ECS (environmental control system) refers to heating and cooling of the cabin interior as well as the installed equipment. Rate gyros measure variations in movement (roll, yaw, pitch axes) during flight. The main bus ammeter showed the pilot electrical-supply amperage levels. The suit flow lever in bottom photo adjusted the airflow rate through an astronaut's suit; the other lever for pilot is hidden by the black attitude control handle. Other items noted above are self-explanatory. Mention of a computer in the Gemini spacecraft previously is labeled here and briefly explained. The MDR shown in top picture above is used to display two-digit addresses and five-digit messages entered into the computer manually through the MDK or read from the computer, to be interpreted by the crew. The MDK was used to manually punch in numbers to read data from the computer by just entering two-digit codes, or, if entering data was desired, a two-digit code plus a five-digit message was also entered. Modes of flight the computer was preprogrammed for, such as launch, catch-up, rendezvous, or reentry, were set on the computer panel or the flight director controller, both of which are pointed out in the bottom photo of the Gemini VII command pilot position, also at the Air & Space Museum, both taken in July 2013 by the author. Lessons learned from the Mercury Project, plus advances in technology and the increasingly complex Gemini-series maneuvers, mandated the adoption of computer technology, since human precision was never exact enough for some critical phases such as reentry firing of rockets, timed to the second to ensure splashdown at a known spot on the ocean. Missing this could result (and did on some splashdowns) in landing far from recovery forces, risking lives of astronauts or loss of the capsule from prolonged exposure in the water. *Both images, author*

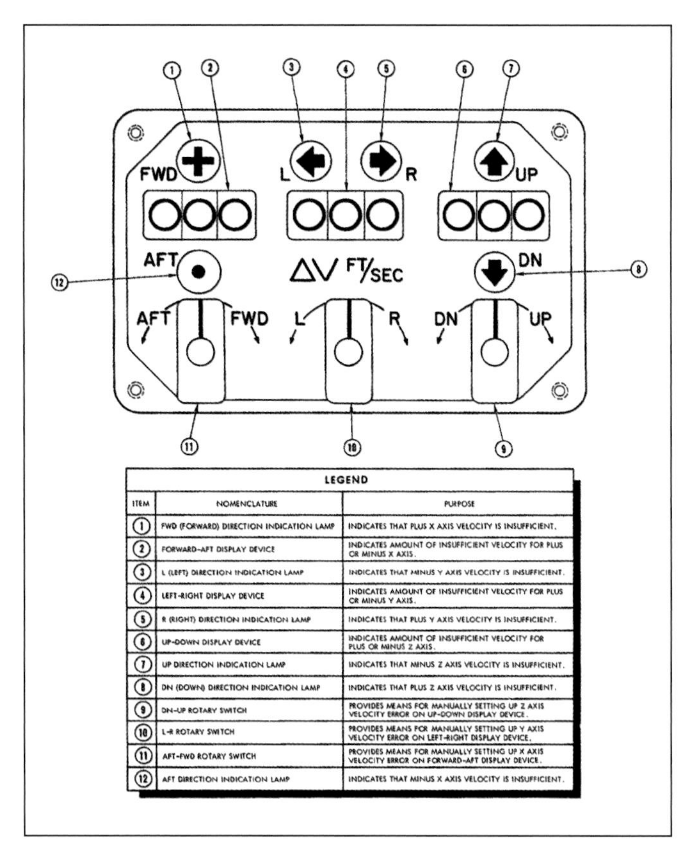

LEGEND		
ITEM	NOMENCLATURE	PURPOSE
1	FWD (FORWARD) DIRECTION INDICATION LAMP	INDICATES THAT PLUS X AXIS VELOCITY IS INSUFFICIENT.
2	FORWARD-AFT DISPLAY DEVICE	INDICATES AMOUNT OF INSUFFICIENT VELOCITY FOR PLUS OR MINUS X AXIS.
3	L (LEFT) DIRECTION INDICATION LAMP	INDICATES THAT MINUS Y AXIS VELOCITY IS INSUFFICIENT.
4	LEFT-RIGHT DISPLAY DEVICE	INDICATES AMOUNT OF INSUFFICIENT VELOCITY FOR PLUS OR MINUS Y AXIS.
5	R (RIGHT) DIRECTION INDICATION LAMP	INDICATES THAT PLUS Y AXIS VELOCITY IS INSUFFICIENT.
6	UP-DOWN DISPLAY DEVICE	INDICATES AMOUNT OF INSUFFICIENT VELOCITY FOR PLUS OR MINUS Z AXIS.
7	UP DIRECTION INDICATION LAMP	INDICATES THAT MINUS Z AXIS VELOCITY IS INSUFFICIENT.
8	DN (DOWN) DIRECTION INDICATION LAMP	INDICATES THAT PLUS Z AXIS VELOCITY IS INSUFFICIENT.
9	DN-UP ROTARY SWITCH	PROVIDES MEANS FOR MANUALLY SETTING UP Z AXIS VELOCITY ERROR ON UP-DOWN DISPLAY DEVICE.
10	L-R ROTARY SWITCH	PROVIDES MEANS FOR MANUALLY SETTING UP Y AXIS VELOCITY ERROR ON LEFT-RIGHT DISPLAY DEVICE.
11	AFT-FWD ROTARY SWITCH	PROVIDES MEANS FOR MANUALLY SETTING UP X AXIS VELOCITY ERROR ON FORWARD-AFT DISPLAY DEVICE.
12	AFT DIRECTION INDICATION LAMP	INDICATES THAT MINUS X AXIS VELOCITY IS INSUFFICIENT.

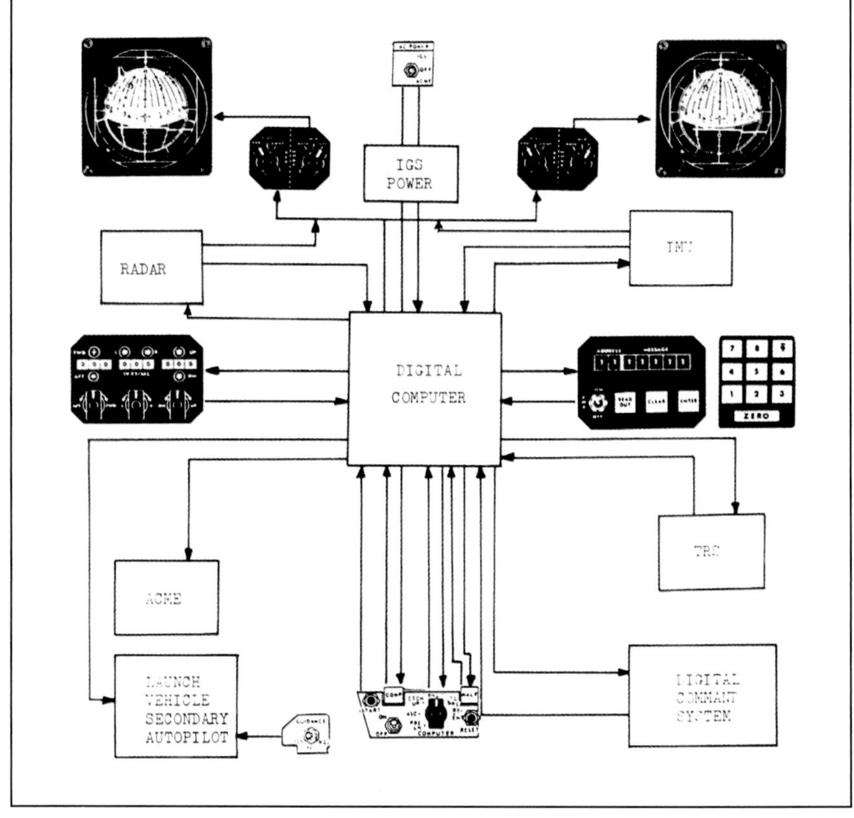

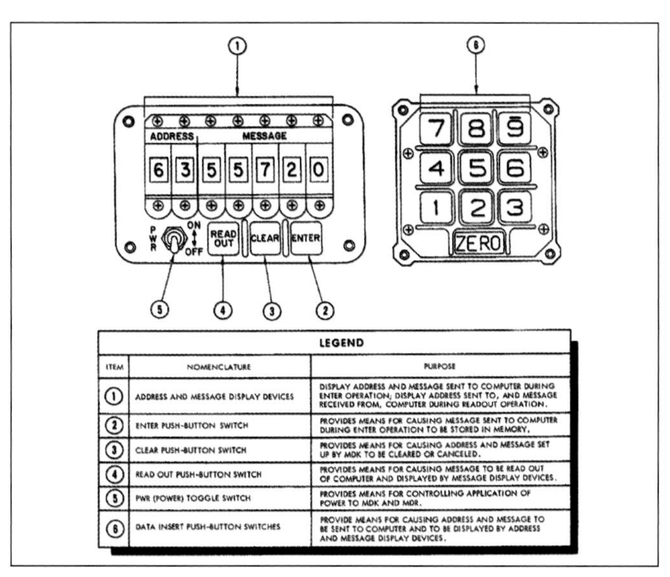

LEGEND		
ITEM	NOMENCLATURE	PURPOSE
1	ADDRESS AND MESSAGE DISPLAY DEVICES	DISPLAY ADDRESS AND MESSAGE SENT TO COMPUTER DURING ENTER OPERATION; DISPLAY ADDRESS SENT TO, AND MESSAGE RECEIVED FROM, COMPUTER DURING READOUT OPERATION.
2	ENTER PUSH-BUTTON SWITCH	PROVIDES MEANS FOR CAUSING MESSAGE SENT TO COMPUTER DURING ENTER OPERATION TO BE STORED IN MEMORY.
3	CLEAR PUSH-BUTTON SWITCH	PROVIDES MEANS FOR CAUSING ADDRESS AND MESSAGE SET UP BY MDK TO BE CLEARED OR CANCELED.
4	READ OUT PUSH-BUTTON SWITCH	PROVIDES MEANS FOR CAUSING MESSAGE TO BE READ OUT OF COMPUTER AND DISPLAYED BY MESSAGE DISPLAY DEVICES.
5	PWR (POWER) TOGGLE SWITCH	PROVIDES MEANS FOR CONTROLLING APPLICATION OF POWER TO MDK AND MDR.
6	DATA INSERT PUSH-BUTTON SWITCHES	PROVIDE MEANS FOR CAUSING ADDRESS AND MESSAGE TO BE SENT TO COMPUTER AND TO BE DISPLAYED BY ADDRESS AND MESSAGE DISPLAY DEVICES.

Major components of the Gemini guidance computer system are shown in NASA drawings to provide clarity. The command pilot IVI (incremental velocity indicator) diagram shown at top left displays values for all three axes (forward velocity, left/right, up/down) in numbers from 000 to 999. The numbers change as any of the three values increase or decrease. Illuminating lamps (1, 3, 5, 7, 8, and 12) show, at a glance, capsule attitude. At bottom left are the two items described on the previous page: the pilot MDR and MDK with features legend. The block diagram at top right shows the major components making up the digital computer system; IMU is the inertial measurement unit, TRS is the time reference system, ACME stands for attitude control and maneuver electronics, and IGS is the inertial guidance system. The computer labeled panel at bottom sets the various modes the computer is programmed for during major flight phases from ascent to reentry, as mentioned previously. Most of these components were located outside the crew compartment, in environmentally sealed enclosures, and accessible only on the ground by removing exterior panels. All systems were run through their paces and passed as ready for flight before being committed to a mission. Despite all of this sophistication, a failure was cause for concern but not considered critical to mission success, since the astronauts were trained in simulator drills on how to compensate by manually setting parameters for flight regimes interrupted by an automatic system anomaly.

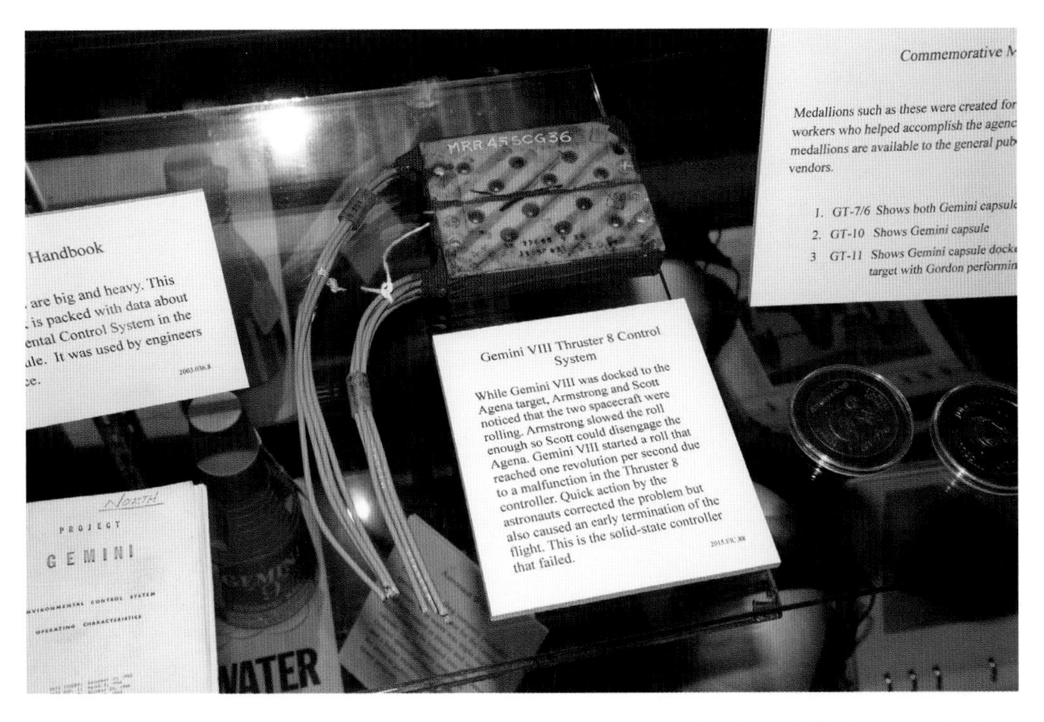

Among Gemini artifacts that have historical significance in particular are those shown in the top picture. The Gemini VIII mission launched Neil Armstrong and David Scott to meet up with an Agena target that had been orbited the same day. They made a successful docking (the world's first), but shortly afterward the mated pair were not controllable. Deciding the Agena must be the faulty craft, they separated from it, only to find that it was the Gemini itself that was now rapidly gyrating due to the number 8 thruster being stuck open, thus bolstering the argument that astronauts have a key role to fill in such situations. The loss of control source was identified and eliminated at the expense of using thrusters reserved for the reentry phase. This resulted in the mission ending early. Even though the docking was not to the desired length of time, a potential disaster was averted, providing valuable experience in space flight. The root cause of the mishap was this item, the solid-state controller shown on display. The bottom image has a publicity pamphlet and water bottle in the shape of a Gemini capsule, distributed to the press corps from the company that supplied purified drinking water to the astronauts. Many similar products of space-related souvenirs were made in the shape of records, coins, movies, books and magazines, flavored drinks, and children's toys. It is still possible to find such things today in such places as flea markets, estate sales, or auctions, and even to find and talk with people who were a part of the space race. In this sense the space hardware category can and does take many shapes and sizes. *American Space Museum Titusville, Florida, by the author*

GEMINI PROGRAM MANNED FLIGHTS

Date	Mission	Crew/Revolutions	Duration
March 23, 1965	Gemini 3	Grissom-Young / 3 orbits	4 hours, 53 minutes; changed orbit 3 times
June 3, 1965	Gemini 4	White-McDivitt / 62 orbits	4 days, 1 hour, 56 minutes; first American spacewalk
August 21, 1965	Gemini 5	Conrad-Cooper / 120 orbits	7 days, 22 hours, 56 minutes; first long-duration flight
December 4, 1965	Gemini 7	Borman-Lovell / 206 orbits	13 days, 18 hours, 35 minutes; rendezvous with Gemini 6
December 15, 1965	Gemini 6	Schirra-Stafford / 16 orbits	rendezvous with Gemini 7
March 16, 1966	Gemini 8	Armstrong-Scott / 6 orbits	first docking; early termination
June 3, 1966	Gemini 9	Stafford-Cernan / 45 orbits	docking failed; Cernan 2-hour spacewalk
July 18, 1966	Gemini 10	Young-Collins / 43 orbits	docking; raised orbit to 475 miles
September 12, 1966	Gemini 11	Conrad-Gordon / 44	orbits docking; created artificial gravity
November 11, 1966	Gemini 12	Lovell-Aldrin / 59 orbits	Aldrin 3 improved EVAs of 5.5 hours

Gemini program total cost: $1,283.4 billion between 1962 and 1968; $10.3 billion in 2018 dollars

To give a better understanding of the role that Agena vehicles played, these NASA images show, *on top*, a boresight test between Gemini 6 and an ATDA (augmented target-docking adapter) in 1966. On bottom is a Gemini 12 photo from 50 feet away of the Agena docking target. The deployed arm contains a dipole antenna with coiled wiring. Black cone is where docking occurs. Green cone is engine exhaust. Docking in space had to be attempted, executed, and experienced by astronauts before the start of the Apollo Project.

Chapter 3

SATELLITES AND SPACE PROBES: CIVIL/MILITARY USES

After Sputnik and Explorer paved the way into low earth orbit, satellites have been populating the near-Earth region in increasing number ever since. They are created, tested, and launched for specific functions and are outfitted with exotic sensors such as high-resolution cameras, filters, antennas, and x-ray, ultraviolet, cosmic ray, micrometeoroid, and infrared detectors. They monitor aerosol pollution in the atmosphere, ozone levels, and sea levels; watch weather patterns; and provide timely alerts in zones when hurricane activity peaks in summer months. Others maintain a steady position in geostationary orbit (otherwise known as Clarke orbit) 22,236 miles from the equator. At this distance, the orbital period of a satellite matches the earth's single rotation. These platforms can provide spot coverage of select regions, acting as radio and television relay stations. Some, such as the Hubble Space Telescope, look out into the farthest reaches of the universe, making epic discoveries of planets, black holes, nebulas, and supernovas. Others aim their optics at Earth, monitoring crop yields and how to use land more efficiently for farming, conducting land surveying, and tracking ice migration, and, at a more personal level, for navigating an automobile or supertanker from one location to another. Aircraft take off, land, and calculate in-flight data from the GPS and GLONASS constellations of precision positioning-system satellites.

Space probes to the planets and beyond have amazed and thrilled humankind with pictures and data as they first flew by the moon, the sun, and nearby planets such as Venus and Mars. As technology improved and experience was gained, others went farther out, to Jupiter, Saturn, Uranus, and Neptune. More experiments and piggyback equipment such as compact landers went along as well, with the mother ships capable of long-duration orbital investigations while acting as radio relays for streams of lander imagery and surface conditions. These landers sampled the hostile environments of places such as Mars, the moon, Europa, and Titan, and of asteroids. Some sampled and brought back soil, rocks, and dust for study. Rovers continue to captivate the scientific community for the pictures of vistas on Mars, as well as for the terrain sampling and weather monitoring they report on daily. They drill into rocks, take temperature readings, and move to new locations when ready to sample another site.

Still others have uses that have been kept in a dark closet as black as deep space itself, with tantalizing exposure to these clandestine activities over the decades. These satellites and their supporting infrastructure belong to governments the world over, for the express purpose of watching what other nations are up to. By using optical camera arrays, electronic intercepts, and radar signals, for example, a picture can be assembled of meaningful intelligence that is used to fund military budgets and to monitor growth and movements of armies, navies, and aviation assets. These robotic spies—from low, medium, and high orbits as well as from stationary perches—are able to keep the peace while also giving advance warning of a sudden cross-border invasion or a missile attack from over the North Pole or from other directions. Some satellites have dedicated military radio and imagery relay functions. Others are hardened to continue functioning in the midst of the electromagnetic pulse (EMP effect) that can short out electronic circuitry from an atomic detonation. Still others provide secure, jam-proof communications links between the national command authorities in the United States and military forces deployed throughout the world, as an example. Others track, follow, and examine satellites of other countries. One thing is certain: if and when the next major war erupts, the first indication of aggressive action may well be the blinding or disabling of satellites, denying space—the high ground—from a nation that signals the opening round of hostilities.

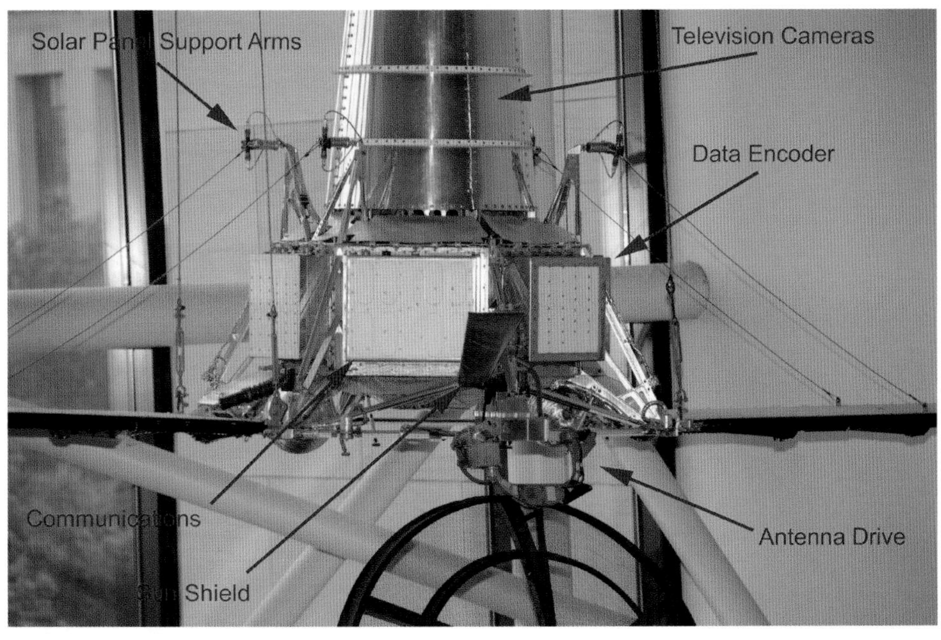

As the first of three steps toward landing men on the moon (the other two being the surveyor soft lander and the lunar orbiter), NASA needed to obtain clear, close-up pictures of the lunar surface. Earthbound telescopes lacked the ability to accurately assess the surface, due to atmospheric interference. To accomplish this objective, a series of spacecraft were built for that purpose. The Ranger probes crashed into the moon intentionally, snapping photos and sending the results to Earth the entire time until impact. Rangers 1 through 6 did not achieve their objectives. A review and overhaul of procedures and equipment configuration (deleting the 94-pound capsule to be ejected before impact to record seismic activity, among other changes) resulted in Ranger 7 proving to be the lucky one, sending back 4,316 images in July 1964. The last photos taken from a mile above the moon showed an area of 100 by 165 feet, so sharp that objects 3 feet across could be viewed. Ranger 8 was able to send back 7,137 photographs before that mission ended in a crash landing in the Sea of Tranquility in February 1965. Ranger 9 ended the series in March 1965, with 5,814 pictures, some with a resolution of objects 1 foot in diameter. It was terminated when it crashed inside the crater Alphonsus. Launching rockets were the Atlas-Agenas. An array of six cameras (two wide-angle, four narrow-field) were enclosed within an aluminum fairing with a pair of solar panels that unfolded after launch. Atop the fairing in the top photo is an omni antenna. A high-gain antenna dish transmitted and received signals. Weight of the probes varied, with some over 600 pounds, increasing to over 800 pounds in later missions. Stability was provided by nitrogen-gas-powered thrusters. Other items of note are seen in the bottom image. The project cost $260 million, and after reviewing all the imagery, it was seen that the moon had plenty of smooth areas, enough to land safely on. *National Air & Space Museum photos, July 2013 by the author*

Mariner 2 replica. It performed the first planetary observation mission to Venus in December 1962, after a 109-day transit. Large round dish is the high-gain antenna. *NASM 7-13 by author*

Mariner 4 replica seen at Kennedy Space Center in September 1988. Launched to Mars on an Atlas-Agena D rocket, it was able to take and send twenty-two images after a 325-million-mile journey. *Author*

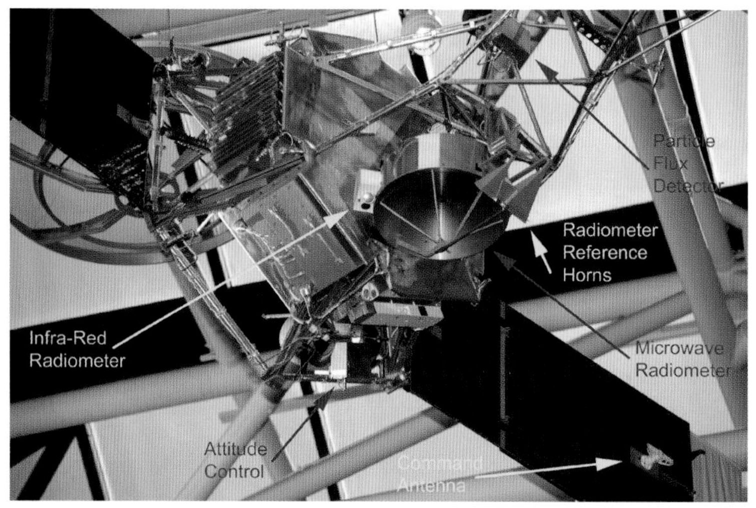

Mariner 2 close-up, with items of note labeled. The basic structure is the same as for the Ranger, with equipment load-out customized for a planetary survey mission. *Author*

Mariners 6 and 7 carried this combination medium/high-resolution camera for taking photographs. It is one of a suite of instruments that spacecraft have installed to maximize data retrieval. *Hazy Center, author*

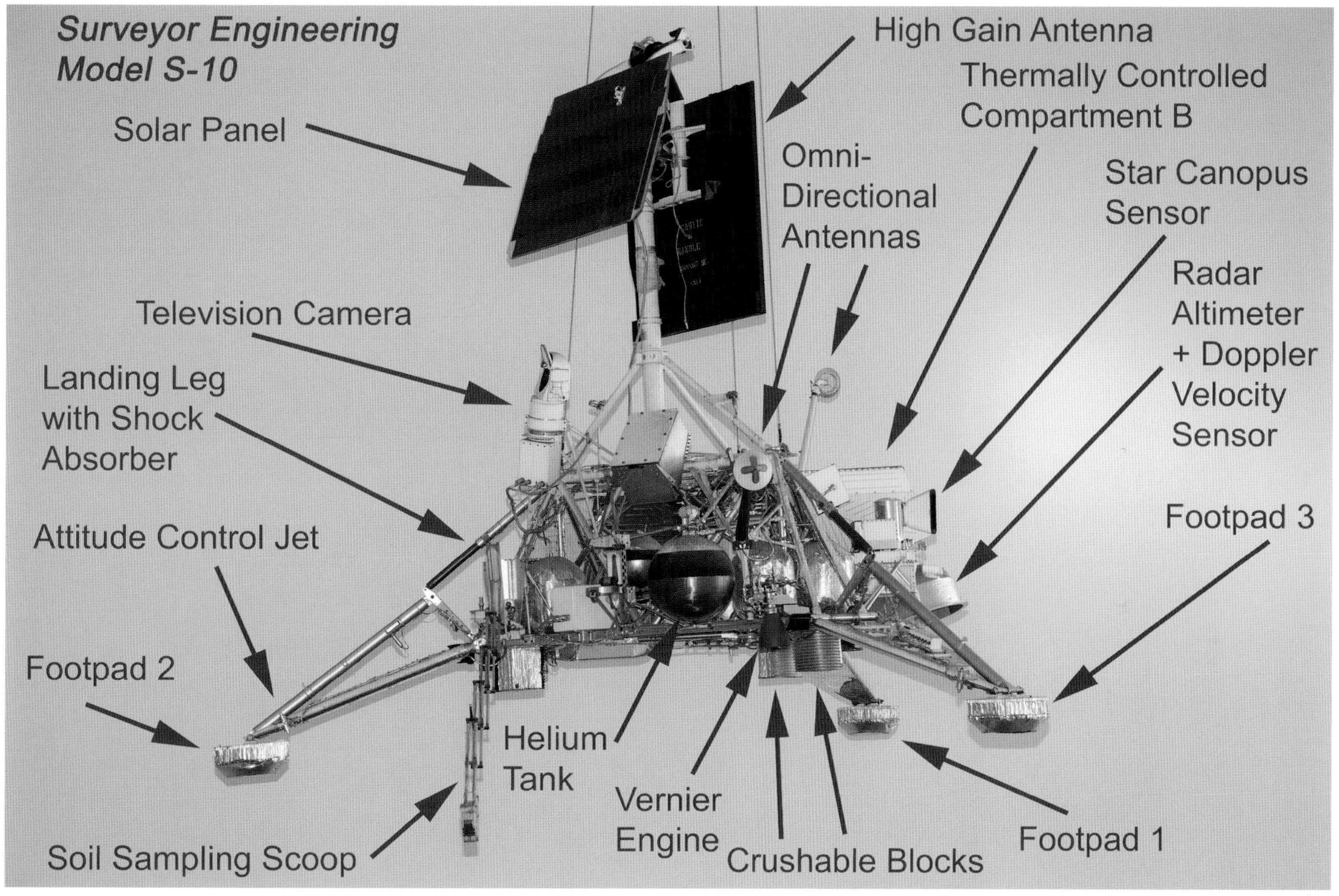

After the Ranger series came the Surveyor soft landers to perform landing and sampling operations on the lunar surface in preparation for the Apollo manned missions. Unknown were facts such as could a heavy vehicle land and be stable on the moon? Could a sampling device dig into the soil? What was the soil composed of? The labeled item above is Surveyor S-10, the engineering model suspended in the air by cables in a gallery at the National Air & Space Museum in Washington, DC; this photo taken in July 2013 by the author.

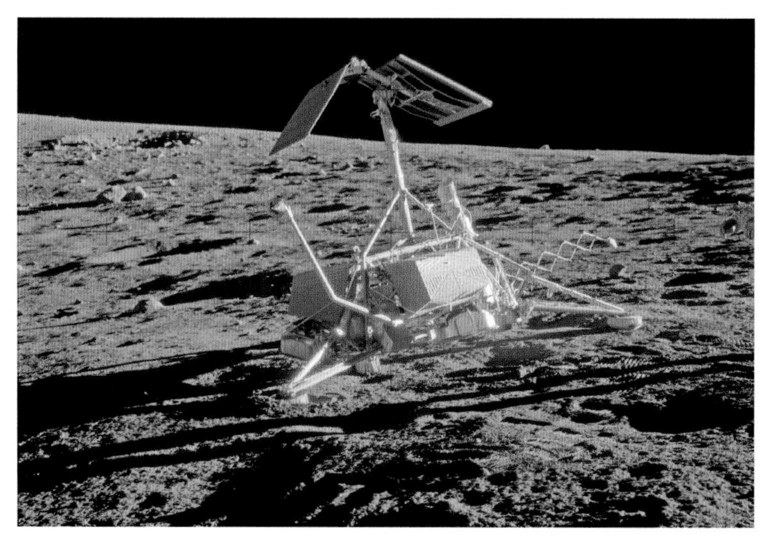
Surveyor 3 bounced on landing and ended up on the edge of a small crater, where it's sat from April 1967 until today. The Apollo 12 crew shot this and a few other photos before removing selected parts from it. *NASA photo*

Gold box containing the Alpha Scattering experiment to determine atomic composition of the lunar regolith. It was attached by cable to the white arm above it. Round base folded away so the box could be lowered onto the surface. *Author*

At this angle, one can see the large white container used to house instruments requiring temperature control. Gold-plated manifold with green nozzle is one of the Vernier engines. Silver component below container is radar related. *Author*

Above image was taken in 2002. By 2013, the Alpha Scattering device had been removed. Ribbon electrical connector is visible. Blue scissor-armed object is extended sampling scoop. White box behind it is an auxiliary battery. *NASM by author*

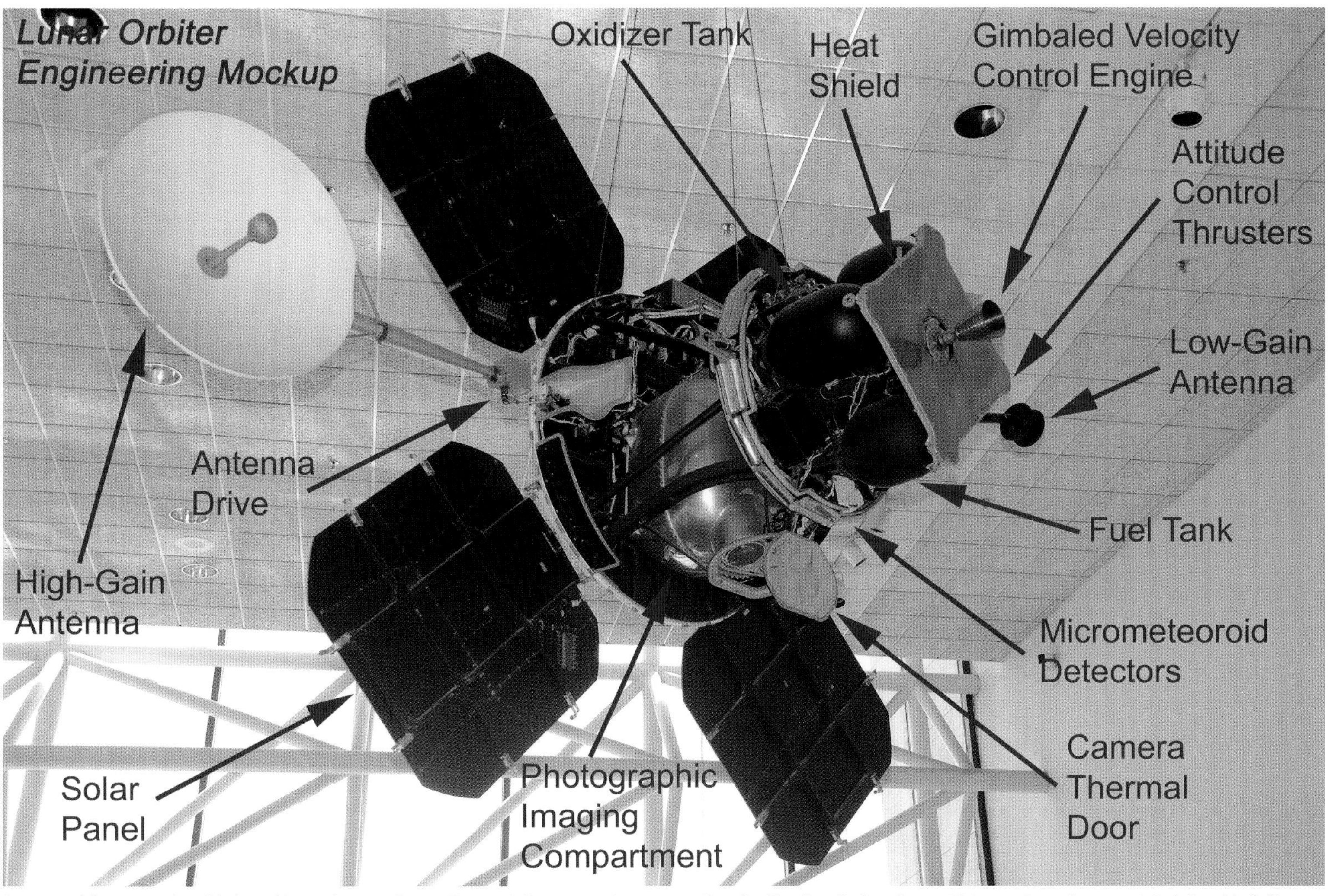

Lunar orbiter was the third and last phase of scouting out the moon in preparation for the Apollo landings. This engineering model at the Air & Space Museum has items of interest labeled. These spacecraft orbited the moon with an emphasis on photographing potential landing sites within the equatorial region from 43 degrees east to 56 degrees west. Five lunar orbiters took pictures from 1966 to 1967 at altitudes with a perigee (perilune or periapsis) as low as 31 miles, resulting in viewing surface details as small as 1 meter across. The climate-controlled photographic enclosure contained an 88-pound camera system with 24-inch (high-resolution) and 3-inch (wide-angle/medium-resolution) focal-length lenses. Both apertures were set at f/5.6. Shutter speeds available were 1/23rd, 1/50th, and 1/100th of a second. Exposed film was chemically processed, the images were scanned, and the data stream was broadcast to Earth, a single image every forty minutes. Irregularities in their orbits led to the discovery of mass concentrations (mascons) of regions primarily where the lunar "seas" existed, where gravitational forces were excessive. The ability to take from 194 to 212 pictures was possible on a film reel length of 200 feet. *Author*

This lunar orbiter view shows, *to left*, the rocket engine. Propellant tanks, an array of micrometeoroid detectors, and a camera section with solar panels are below. Canopus star tracker is above the camera bay. *Author*

The thermal door above the camera lenses was part of the temperature regulation procedures used to avoid excessive heating or cooling of the photographic instrumentation while orbiting the moon. *Author*

In this view, at left is the fairing for the deployed silver mast of the high-gain antenna. For launch it was folded against the body of the orbiter. Black box at center is an electronic switching device. *Author*

In another design feature to control extremes in temperatures, the back shield was painted white to reduce the effects of solar heating. A sun sensor used to orient the spacecraft is the round aperture at the nine o'clock position. *Author*

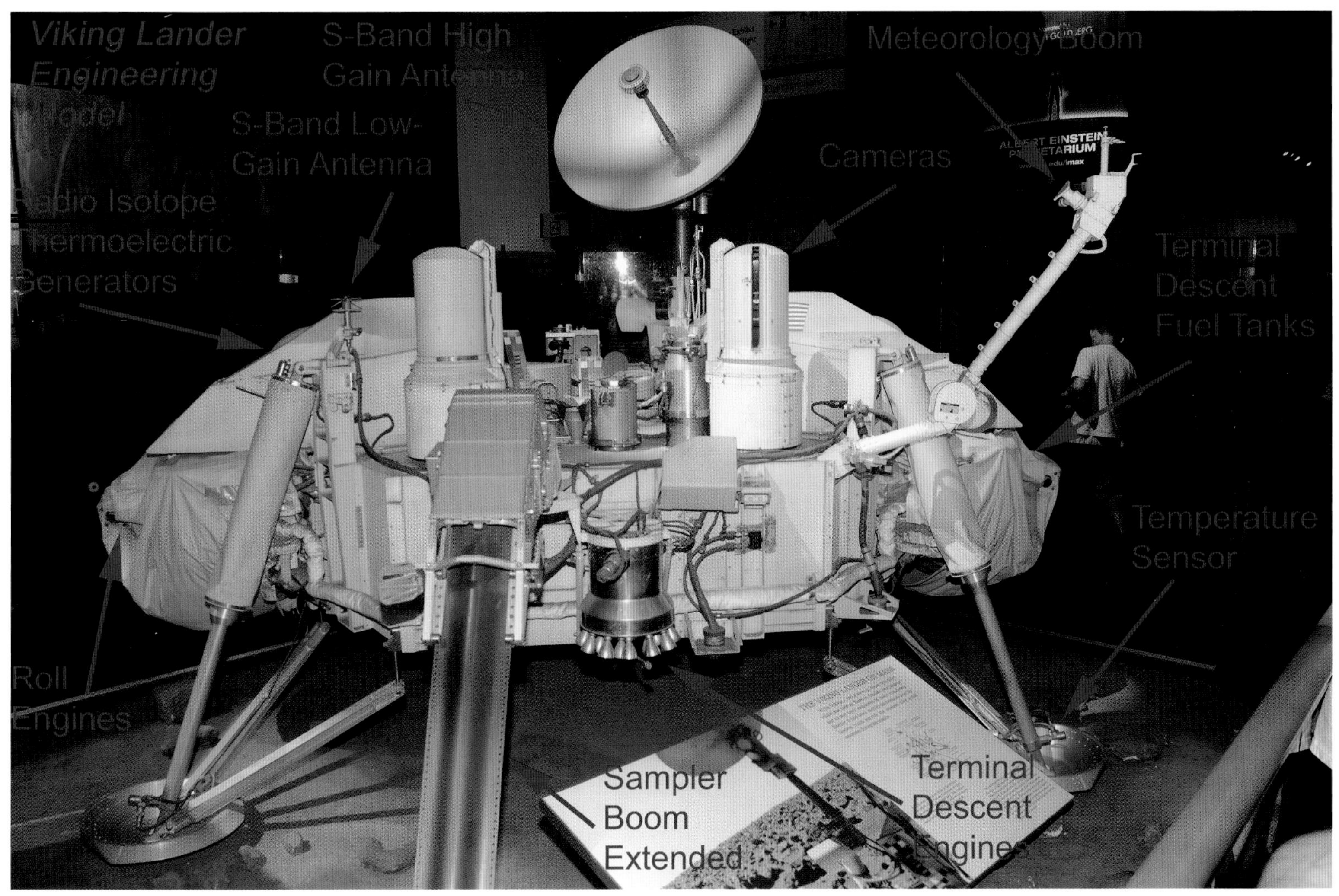

This Viking lander at the National Air & Space Museum was used as a test bed during actual Mars operations. It was encapsulated underneath the Viking orbiter probe, being released when at Mars to conduct surface investigations. Inner structure was a hexagonal (six-sided) box 18.2 inches deep. Height from dish antenna to footpads was 7 feet. Three terminal-descent engine arrays fired at 585 pounds of thrust each. Nozzles (eighteen per unit) were used to dissipate the exhaust plumes, to avoid intense exhaust blast effects at the landing site. Power was provided by a pair of systems for nuclear auxiliary power (SNAP-19) radio-isotope thermoelectric generators (RTGs). Power output at the beginning of lander sampling was 42.6 watts. Lander weight was 2,353 pounds. *Photograph taken by the author in July 2013*

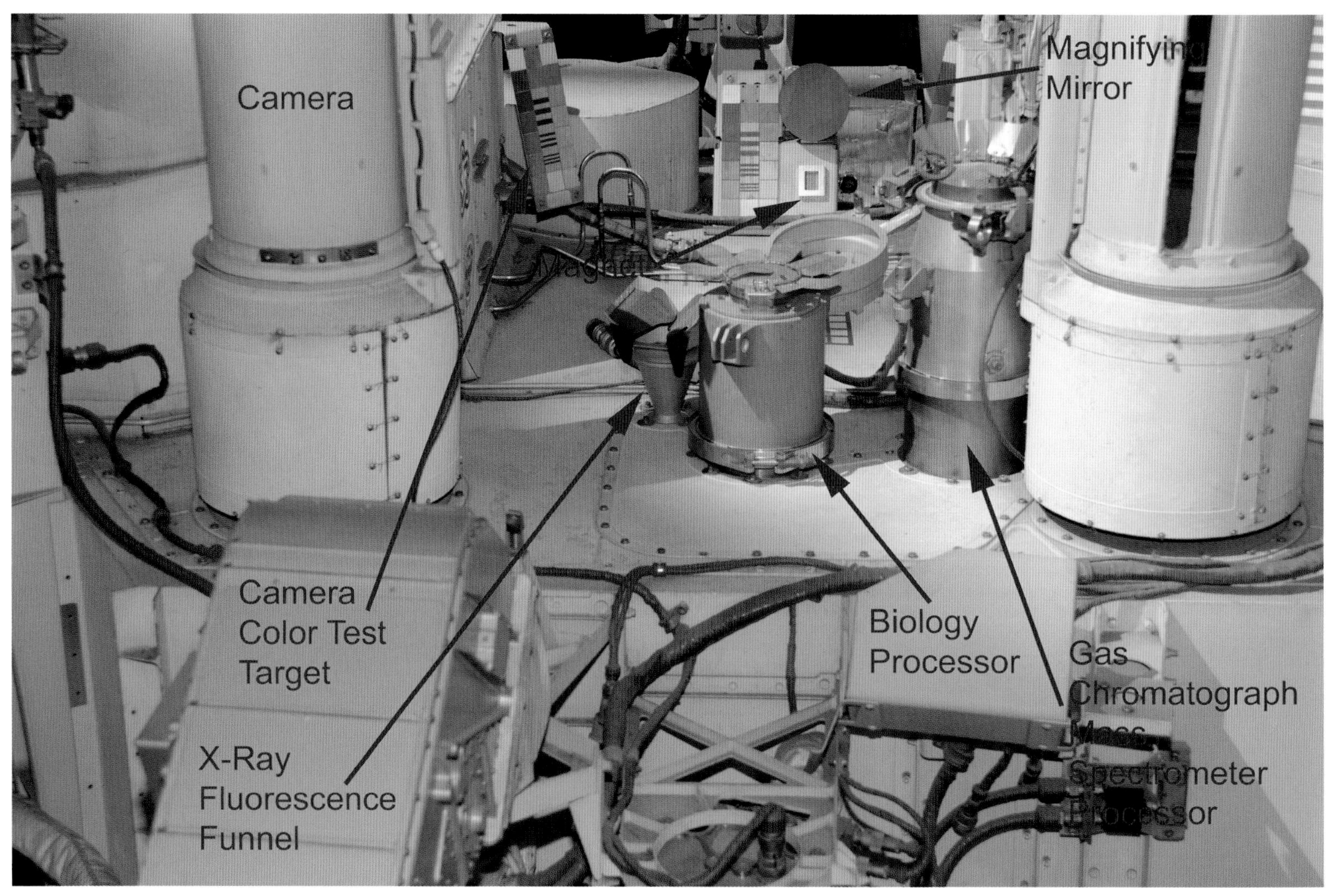

The pair of Viking landers (Vikings 1 and 2) were equipped to conduct comprehensive testing of Martian soil samples to determine if life was present. The x-ray fluorescence device performed inorganic assessments. The biology processor diagnosis results centered on photosynthesis, growth, and metabolism if present. The last funnel accepting soil into its funnel from the sampler scoop was the gas chromatograph mass spectrometer processor or GCMS, focused on organic compounds that could be found. No conclusive evidence was proven from either lander analysis. This view is of the top deck section of the National Air & Space Museum lander on the previous page. Gold component behind the magnifying mirror is the radar altimeter electronics. *Author*

View from a reverse angle shows the seismometer (*at center*). They are used to detect, record, and transmit vibrations associated with Mars quakes, tremors, or volcanic activity. *Author*

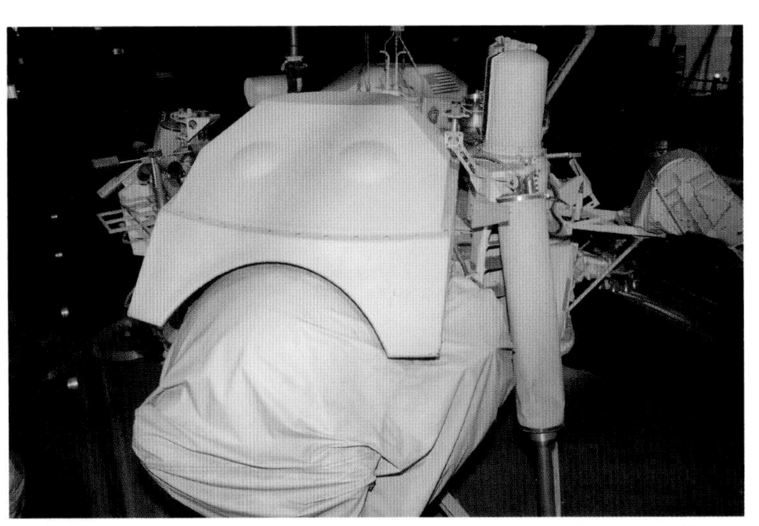
The RTG cover is the hard shell above the round tank of propellant liquid, covered in insulation. An identical configuration is at the opposite end of the lander. UHF and S-band antenna flank cover. *Author*

Viking lander UHF antenna array. As the take of data was compiled, messages of findings were relayed to the orbiting Viking platform for retransmission to Earth. The seismometer unit is at upper left, RTG cover to right. *Author*

At the Seattle Museum of Flight, Viking 3 is on display, with this view showing an open insulated bay in the foreground where the seismometer would be installed. Two round housings in background are for the cameras. *Author*

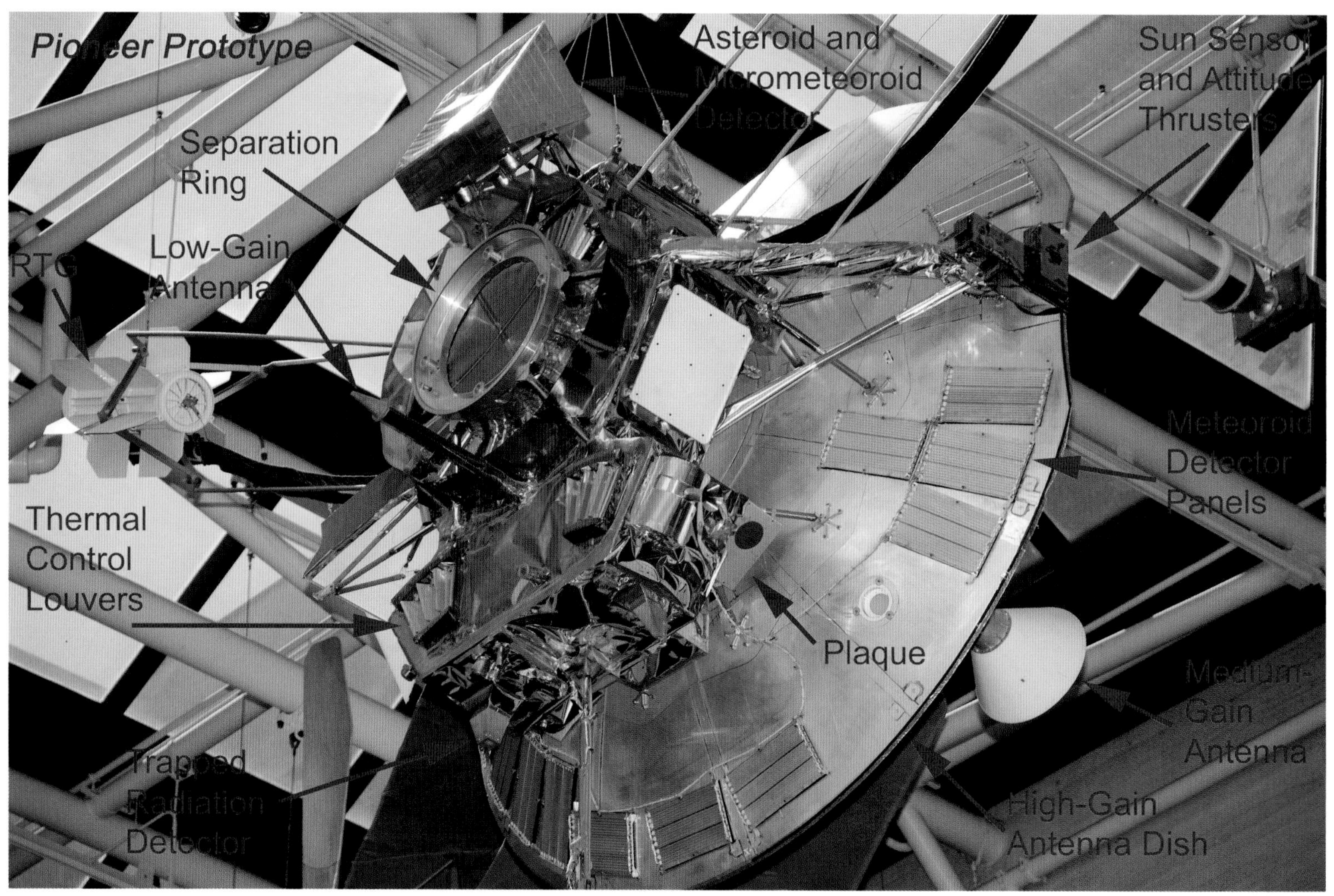

Pioneer prototype shown at the National Air & Space Museum. Highlighted components labeled. Of note is the plaque location, having an outline of the spacecraft. Also illustrated are figures of a man and woman, the third planet in our solar system where the probe originated from, and fourteen lines of radiated energy originating from various pulsars, with our sun at the center. A hydrogen atom symbol is also included. Pioneer 10 is on a two-million-year trip to the star Aldebaran (the eye of the constellation Taurus). Its last signal sent to Earth took eleven hours, twenty minutes to arrive. Pioneer 11 will approach the star Lambda Aquila in about four million years. *Photo taken by author*

Pioneer high-gain antenna is 9 feet in diameter, with the medium-gain, cone-shaped receiver supported by three struts. The slot at lower center is for the plasma probe aperture. Spacecraft weighed 570 pounds.

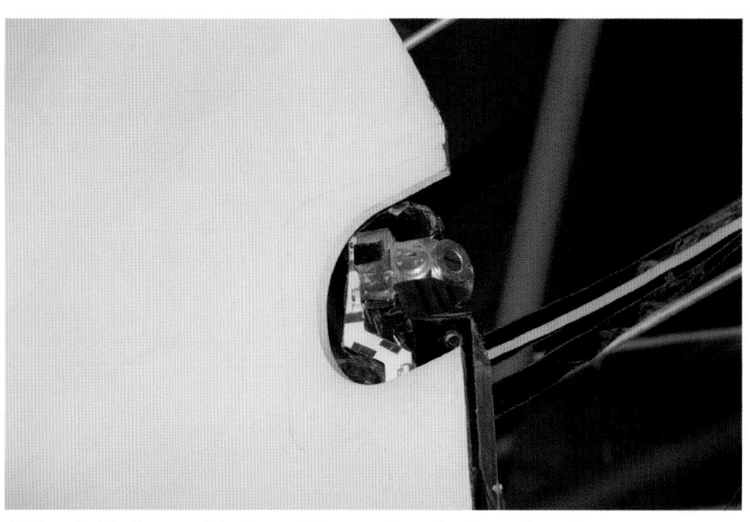

At the dish's three o'clock position are two devices: the silver sun sensors to orient the platform correctly in space, and, below this, a foil-wrapped attitude thruster housing. An array of these is as important to success as the experiments themselves.

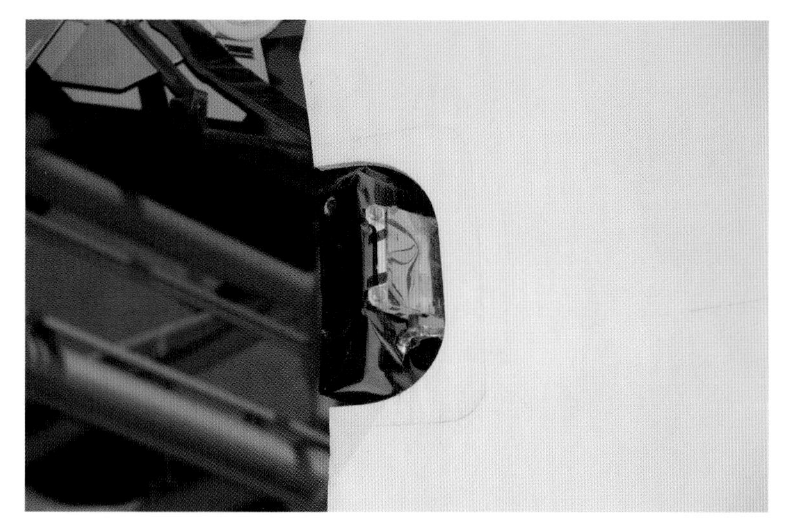

At the nine o'clock position of the edge of the dish is this notch cutout to enable mounting of a set of attitude thrusters that fire when minute corrections need to be made. Adjustments become critical when passing a planet or moons to be studied.

Two booms extended after launch to move the RTG power sources 10 feet away from the central section of instruments. Cables route electrical energy to spacecraft systems. Fins are used to radiate excess heat into space. *Photos this page by the author*

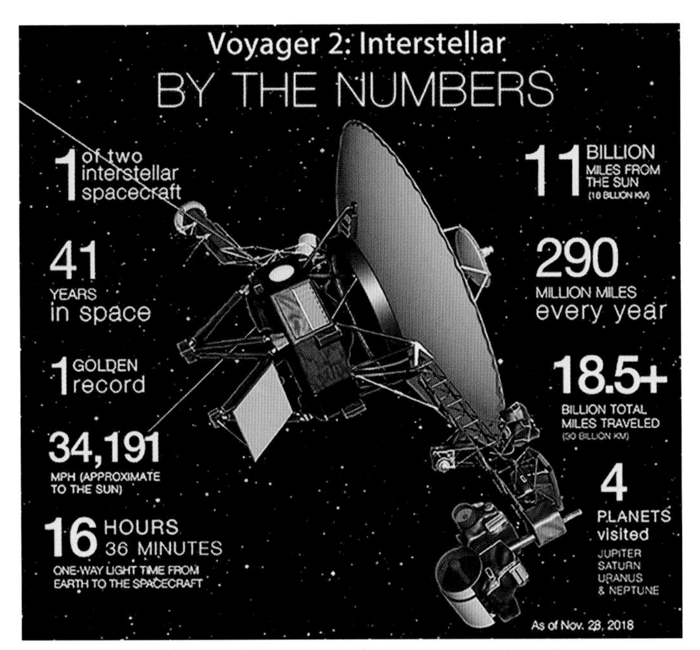

Voyager space probe statistics as of November 2018. Both were launched to study Jupiter, Saturn, and the outer solar system environment. They also discovered twenty-two new moons of four planets. *NASA*

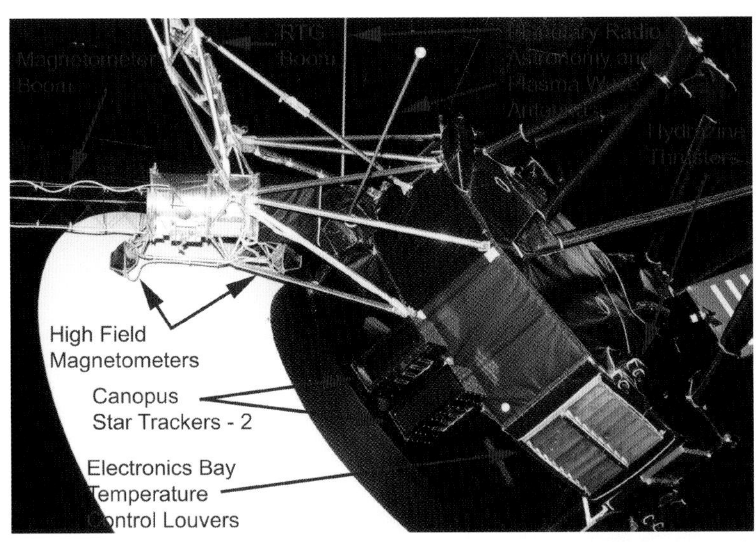

View of Voyager DTM (development test model) at the National Air & Space Museum. The RTG and magnetometer booms deployed after the probe was free of the rocket. The highlighted section is behind the main dish antenna. *Author*

The scan platform boom in deployed position. It is articulated to move cameras and sensors, keeping objects centered for clear photographs. This prevents the smearing effect of images that can occur. *Author*

The white high-gain antenna has a black sun sensor aperture above the low-gain antenna support arms. Flat section of black electronics bay above the dual star trackers is where the gold-plated record would be attached.

The Sounds of Earth
Gold-plated copper record is a duplicate of the ones on Voyagers 1 and 2 as they move ever farther out into deep space. It is 12 inches in diameter and contains pictures and various animal sounds, including whales and birds. Also included are greetings by humans in fifty-five languages, as well as music. It is inside an aluminum case with instructions and a stylus for playing should intelligent life forms come into its possession. *Hazy Center, July 2013 by the author*

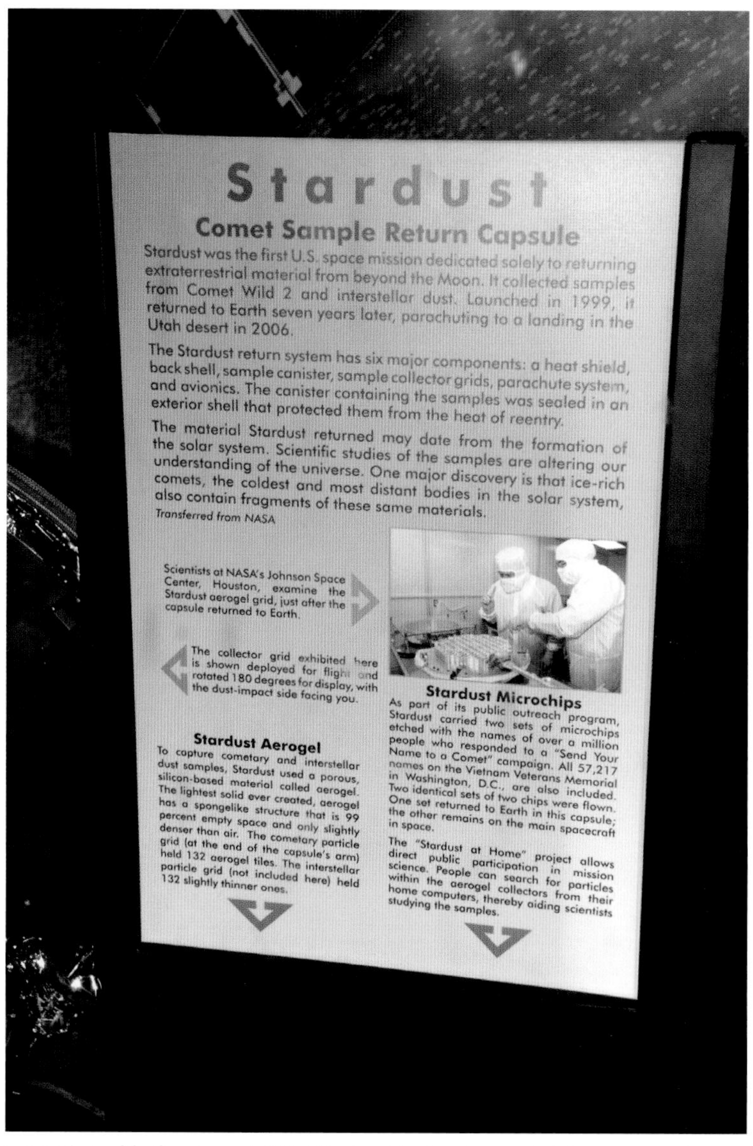

The Stardust mission launched a probe in 1999 on a trajectory to a close encounter with the comet Wild 2. A grid of cells containing aerogel were exposed to the stream of small fragments emanating from the comet's tail, then sealed inside a reentry capsule that landed on Earth in 2006. Some of the grid cells were missing while being examined by scientists. NASA also sponsored a "Send Your Name to a Comet" website where people could have their names loaded onto a microchip that went on the mission. The author entered his name, along with over a million others who responded to the offer. *Air & Space Museum photos by the author*

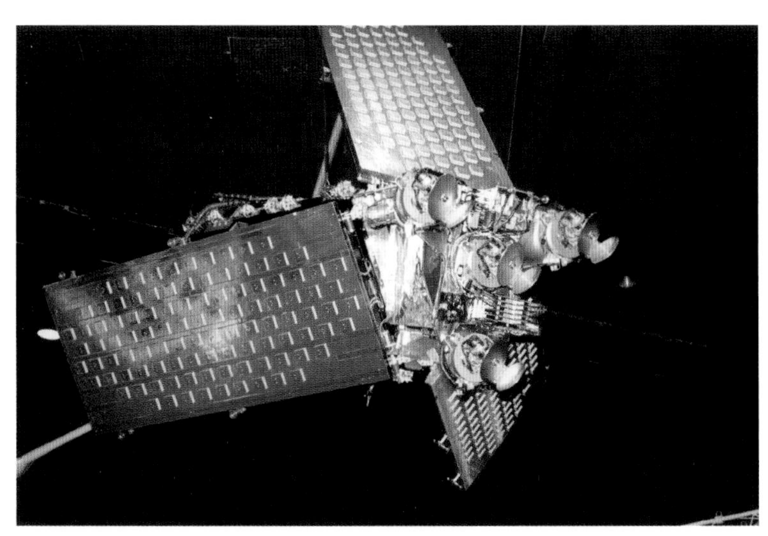

An Iridium satellite, one of a constellation of sixty-six used for cell phone services. The three angled panels are L- + S-band antennas, each with 106 radiating elements. Center array has gateway/cross-link antennas. *Author*

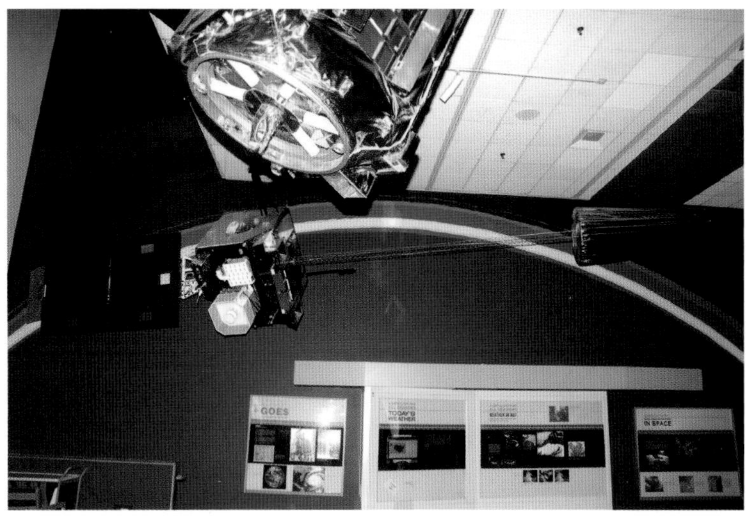

GOES: Geostationary Operational Environmental Satellite half-scale model. Solar sail at right is folded; solar panel to left has a trim tab at very end. Central body is where all of the sensing and reception/transmissions occur. *Author*

This angle shows the dual solar panels mounted at the aft end. Iridium 33 collided with Kosmos 2251 on February 10, 2009, at a speed of 26,000 miles per hour. It was the first hypervelocity satellite impact. *Both NASM by author*

GOES sensors: large white UHF antenna; slotted white one above is S-band receive. Pole antenna between them is for SAR reception. Upper left section with two apertures are earth sensors. Black hoods are for imager and sounder. *Author*

ITOS-1: improved TIROS operational system prototype at the National Air & Space Museum. This view shows the round momentum wheel at base, along with two very high-resolution radiometers. *Author*

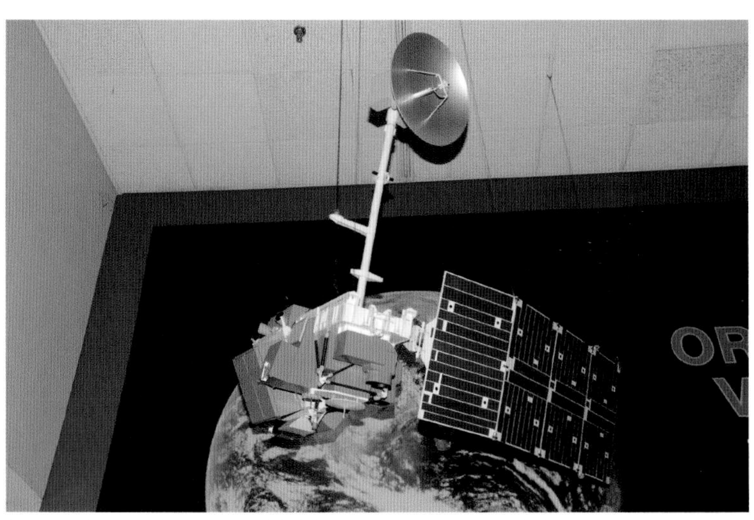
Landsat 4 model at NASM. Mast has high-gain dish antenna. On mast is small arm pointing left with GPS antenna. Blue component in foreground is a multispectral scanner. Gray angled modules are power distribution. *Author*

ITOS-1 sensors include S-band antenna inside black circular device at top; pairs of round optics along both sides are vertical temperature profile radiometers. Bottom arrays are digital solar intercept sensors. *Author*

Multispectral scanner (MSS) unit used on the Landsat earth-orbiting satellites. Images cover 13,000 square miles with a resolution of 250 feet. They document changes in land usage over time. *Photos NASM except lower right by author*

The Sirius FM-4 satellite displayed at the Hazy Center was a flight spare donation. It represents the first generation of satellite radio platforms providing twenty-four-hour broadcast capabilities. *Author*

The Wilkinson Microwave Anisotropy Probe, or WMAP, was able, over a nine-year period, to map the cosmic microwave background. Mapping cycles took six months with measurements as precise as 2/10,000th of a degree. *NASM by the author*

On top of the Sirius FM-4 satellite are seen various items, such as thrusters to maintain geostationary orbit position, antennas, feed horns to direct data streams, and folded solar panels on sides, which if opened would span 78 feet. *Author*

A detail picture of the central structure on the WMAP full-scale reconstruction article. Once all of the microwave analysis had been compiled over the course of this project, the universe was determined to be 13.77 billion years old. *Author*

The Hubble Space Telescope is one project that most of the public can identify and knows of the results it has provided—stunning views of galaxies, novas, interstellar gas, and dust, and even detailed photos of our solar system's planets. The left photo was taken inside the gallery of the National Air & Space Museum in July 2013, showing the structural dynamic test vehicle not intended for use in space. Solar panels are furled along the sides of the light shield of the upper body section. Stowed black dish antenna is for high-gain data use. On the right is a full-scale mockup shown at the Kennedy Space Center *Atlantis* exhibit, with the light-shield aperture door opened and solar panels deployed. Hubble was designed for servicing in orbit by the space shuttle. After launch, a defect was found in the ability to obtain sharp focus on objects of interest. A shuttle mission installed corrective optics, fixing the problem. Future visits kept Hubble healthy and relevant, with new capabilities. *Both photos, author*

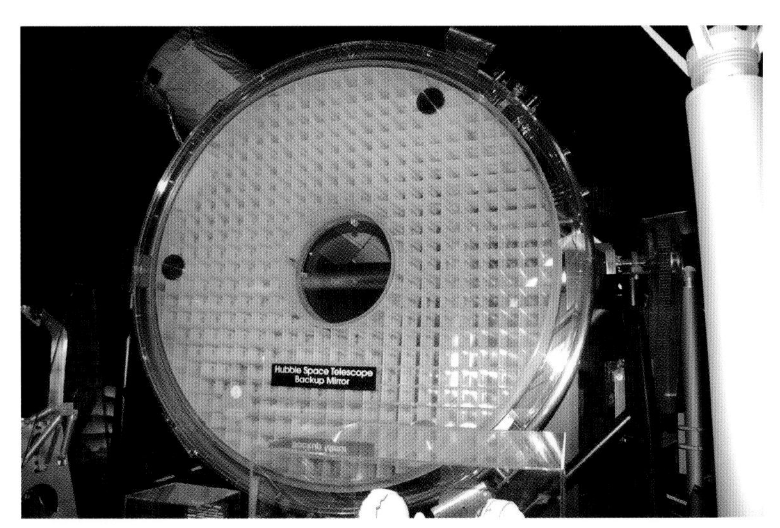

Hubble primary mirror spare. Diameter is 7.9 feet (2.4 m). Light is focused on it by a secondary mirror 12 inches (0.3 m) wide, weighing 27.4 pounds (12.3 kg). *NASM by the author*

Handrails and access doors allow easy servicing operations once a shuttle retrieves Hubble for maintenance. This view shows the section where the faint-object camera (FOC) was installed. *Author*

Four labeled access bays are, *from left*, the DIU (data interface unit), OCE (optical control electronics), FGE (fine guidance electronics), and ACE (actuator control electronics). Also seen are two arm grapple fixtures for the space shuttle remote manipulator arm. *Author*

Hubble has three fixed-head star trackers, shown here. They view patterns of stars after orbital attitude adjustments, which the onboard processors use to aim the telescope to an accuracy as fine as 60 arc seconds. *KSC exhibit by the author*

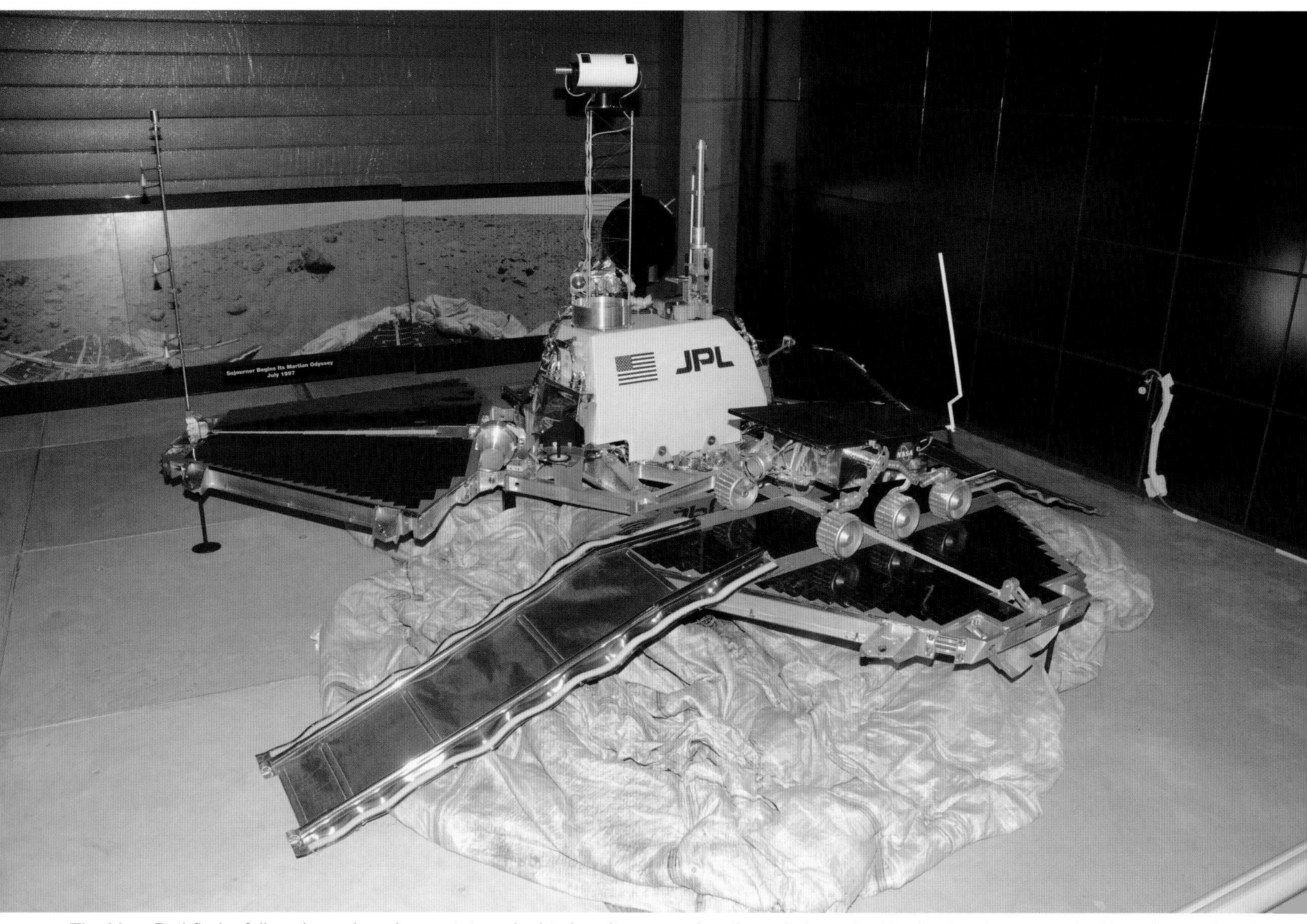

The Mars Pathfinder full-scale engineering prototype depicts how it appeared on the red planet, in this simulated scene at the Hazy Center. Lander and rover were enclosed within the three triangular solar-cell-covered petals, the airbags inflating before touchdown. Bouncing on the surface before coming to rest, the petals opened, exposing the equipment for the start of surface studies. A weather station mast with three windsocks is at left. The white cylinder above electronics housing is the imager for Mars Pathfinder camera. Silver mast to right is a low-gain antenna. High-gain antenna is the black disc in the background. Rover drove down ramp to begin survey of surrounding areas. At front of it is round silver Alpha Proton X-Ray Spectrometer (APXS), used to obtain readings of chemical composition from rocks. Rover (called Sojourner) had solar cells, and a six-wheel drive with heated interior components. Landing was on July 4, 1997. *Author*

The Advanced Orbiting Solar Observatory (AOSO) was a proposed platform for continuous viewing of the sun from visual to x-ray wavelengths from a polar orbit. Canceled in 1965, it later flew on the Skylab. *Hazy Center by author*

Clementine performed a detailed survey of the moon in 1994. In 1996, it was announced that data it gathered showed the presence of water ice at the southern polar region. The example at NASM is an engineering model. *Author*

A tracking and data relay satellite (TDRS) at the Hazy Center. It has two solar panels and a pair of round K-band antennas. The four white items, *left to right*, are a solar sail, omni antenna, K band, and space ground link. *Author*

The New Horizons probe left Earth in January 2006, passing by Pluto in July 2015. It also investigated Kuiper belt object 2014 MU69 / Ultima Thule (officially named Arrokoth) on New Year's Day 2019. It weighs 1,054 pounds. *Hazy Center by author*

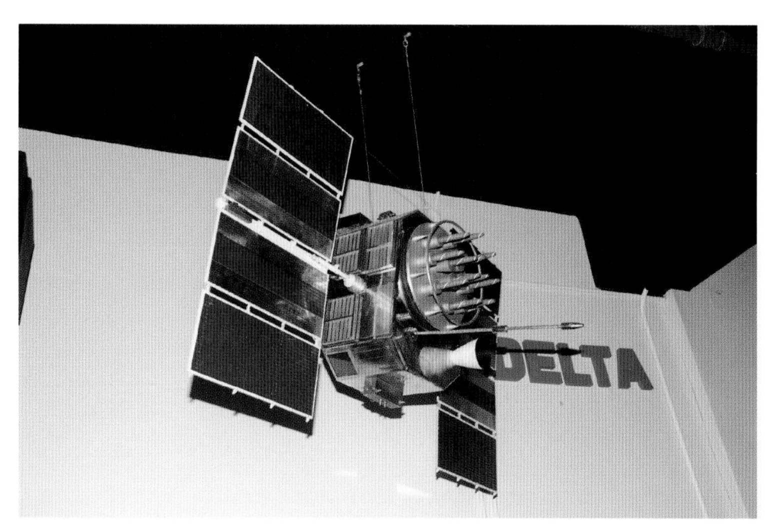

A model of a global positioning system (GPS) satellite. These replaced other land-based stations such as OMEGA and LORAN with a space network of more accuracy, covering the entire globe. *KSC by author*

A Picosat on display at the KSC *Atlantis* exhibit. With advances in microcircuitry and miniaturization of components, these devices are taking over many roles formerly the domain of bigger platforms. *Author*

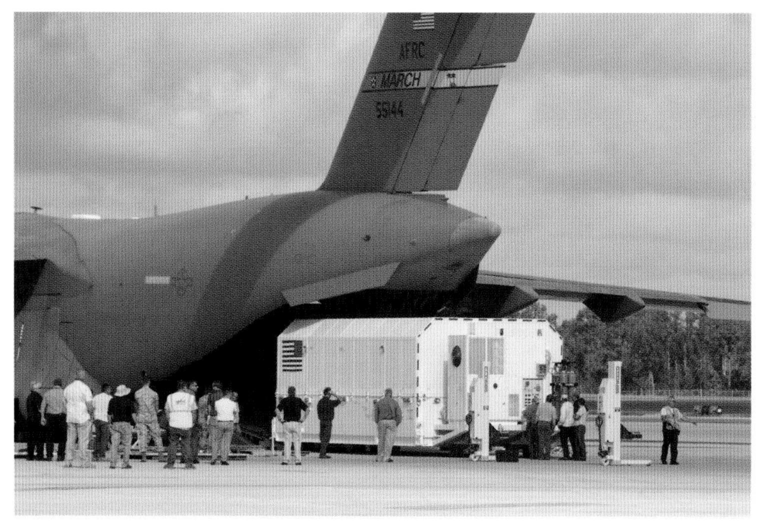

A GPS satellite in transport mode, arriving at the Space Center in Florida enclosed in an environmentally controlled container. Air Force cargo planes, such as the C-17A Globemaster III, handle this task easily. *Author*

With size and weight issues reduced, Cubesats can occupy the nose payload section of a rocket, being released into low earth orbit once the correct altitude and inclination are established. Many nations and companies are exploiting this capability. *KSC by author*

Another early use of space in the 1960s to provide an early-warning method of intercontinental ballistic missile launches was the American MIDAS (missile defense alarm system), which detected hot missile exhaust plumes in the colder atmosphere and space environment by utilizing the infrared bandwidth. This was the first attempt by any nation to orbit such a detection capability. Up to twelve polar-orbiting satellites would provide almost total coverage of the Soviet Union, but in reality fewer than the twelve were ever operational. The system suffered from a high false-alarm rate, caused by reflections of sunlight off clouds and the polar ice caps. Although considered marginally successful, the experience gained and technological advances that resulted were applied later into the successful DSP (defense support program) space platforms. *Left photo, MIDAS internal contents, Cape Canaveral Air Force Station, June 2013; right photo, MIDAS exterior fairing, Hazy Center, July 2013 by the author*

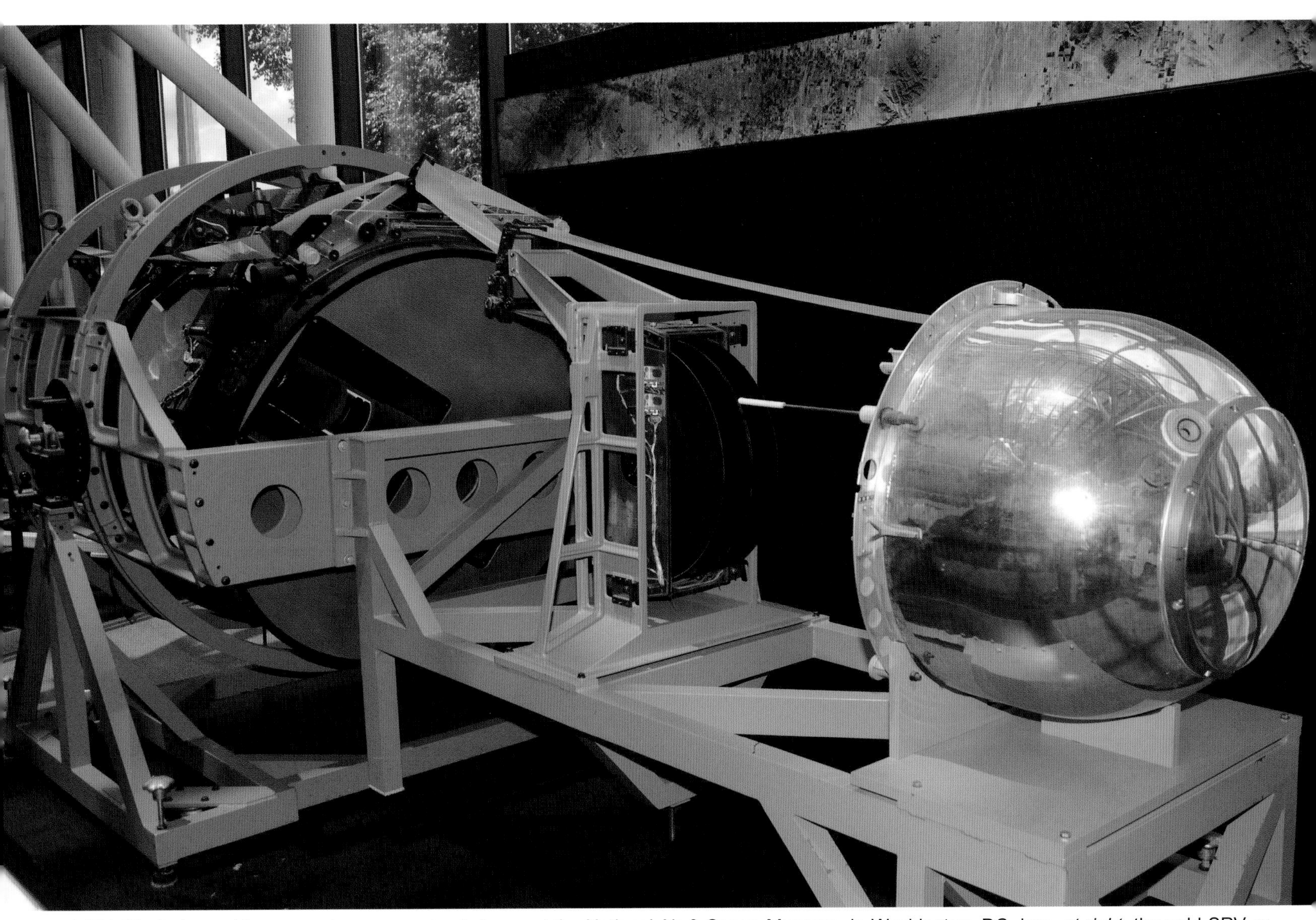

A KH-4B photographic reconnaissance payload shown at the National Air & Space Museum, in Washington, DC, has, *at right*, the gold SRV or satellite recovery vehicle. It brought back to Earth exposed spools of film to be developed and scrutinized by photographic interpreters. In center is where another SRV should be, but only the take-up film spools are present. At left is one of the angled stereo panoramic cameras. Two were used to give interpreters the ability to view identical images in stereoscopic viewfinders. Tan film supply can be seen routed across the top of the mechanical workings, ending up in the forward SRV first. When full, the film was cut and the capsule or "bucket" was ejected for its trip back. The second SRV then was filled until it was also separated and deorbited, to be snatched in midair by an Air Force cargo plane standing by with a retrieval device ready to snag the parachute and haul the SRV in. *Author*

KH-4B forward SRV in 2002, with the gray forebody or heat shield in place. The two rolls of film had identical images on them that were exposed by the pair of stereo cameras. Also missing is the retrorocket motor. *Author*

The supply source of film was at the aft end of the unit. Early film proved brittle in the cold space region, so in its place a polymer-based film was used successfully. Capacity was 36,000 feet of black-and-white film. *Author*

Midsection had the pair of stereo cameras, with a central triangular section or "delta structure" as a fixed part of the interior body. Resolution of the printed images was on average from 5 to 7 feet. *Author*

Main intermediate roller assembly advanced the film as it was exposed. Drive motors, rollers, and guillotines to cut the film—all had to work flawlessly from launch to on-demand operations in low earth orbit. *Author*

Agena-A, *on top*, was used to loft payloads such as Discoverer 1, Midas, and the Samos (satellite and missile observation system), as well as early electronic intelligence or ELINT payloads. It had a gimabled XLR-81-BA-3, or -BA-5, rocket engine that developed 15,500 pounds of thrust for up to 120 seconds. Between 1959 and 1961, twenty were used in total. Agena-B, *on the bottom*, had an upgraded XLR-81-BA-7 engine that was capable of restarting after placement in low earth orbit. This allowed for changes in orbital paths. It was lengthened to accommodate larger fuel tanks, increasing burn time to 240 seconds. Later it was fitted with the XLR-81-BA-9 engine with uprated 16,000 pounds of thrust. Seventy-six of the Agena-B model were used as upper stages. Some sent the Ranger series out of earth orbit on a trajectory to the moon. Also orbited were all the early Corona (KH-2/KH-3, some KH-4 and KH-5 ARGON). Agena-C was a proposed follow-on (increasing fuel tank size to twice that of the -B version), but in any event was never built. Agena-D was the most produced and versatile of the series, becoming the first general-purpose satellite. Used in the NASA Gemini Project as a docking target and launched on Atlas and Thor rockets, it was easily adaptable to many payloads, including those used for photographic reconnaissance, such as some Thor-Agena launches of KH-4, KH-5, and KH-6 Lanyard types. Mated to the Atlas, Agena-D lofted KH-7 Gambit photo-imaging platforms as well as the Mariner space probe. It also was used with the Atlas rocket in placing Canyon, Aquacade, and Rhyolite ELINT-SIGINT-gathering (electronic intelligence / signals intelligence) satellites into geostationary orbits. From 1962 to 1987, 369 of all variants were assembled and flown. Dry weight averaged 4,000 pounds (1,800 kg). Fully fueled and outfitted for a Gemini mission, weight increased to approximately 18,000 pounds (8,200 kg). The Air Force designation was RM-81 (RM = research missile). Fiscal year serial number records, totaling 341 Agena examples produced, began in 1959, ending in 1963. These photos were taken in 1994 at the Cape Canaveral Air Force Station Space Museum. Both these artifacts have since been relocated, completely restored, and enclosed inside a hangar for public viewing. *Both photos, author*

Shown at the Steven F. Udvar-Hazy Center is this Agena-B along with ground station consoles used by engineers and technicians before launch and during orbital missions. With a gimbaled and restartable engine, the Agena was versatile, adaptable, and flexible for many mission profiles, from the launching of early satellites for exploration of the moon and planets on precise trajectories, to payloads designed to observe activities on Earth. A panel removed to left of the USAF marking houses various electronic components. *Author*

This image and the one below are of Discoverer XIII, the first space object successfully recovered by the US Navy from the Pacific Ocean on August 11, 1960. It was a dress rehearsal for future film return flights. *NASM by Author*

This C-119J Flying Boxcar, s/n 51-8037, made the first midair retrieval of an object in space when it reeled in the capsule ejected from the Discoverer XIV satellite over the Pacific Ocean on August 19, 1960. *USAF Museum by author*

No film was returned in the Discoverer XIII capsule; this view shows no spools for film, only recovery and analytical equipment. An American flag was the payload, presented to President Dwight D. Eisenhower. *Author*

It was this capsule, preserved at the USAF Museum, that the C-119J captured on its third pass over the descending object. Inside was exposed film taken over several locations by a KH-4 Corona photoreconnaissance platform. *Author*

Visitors to the National Museum of the United States Air Force in Dayton, Ohio, can view three Cold War–era satellites that operated in space for the purpose of observing activities behind the Iron Curtain. This is a KH-7 Gambit photoreconnaissance non-flight-rated example. The top image shows, *to left*, the gray heat shield / forebody that surrounds the gold capsule/satellite recovery vehicle or SRV. Cone-shaped thrust body contains the retrorocket used to slow the SRV prior to reentry. The white truncated section is the adapter to the aft compartment containing the camera, mirrors, and corrective optics. Silver insulated area with window was aimed at targets of interest. *Both pictures*, *author*

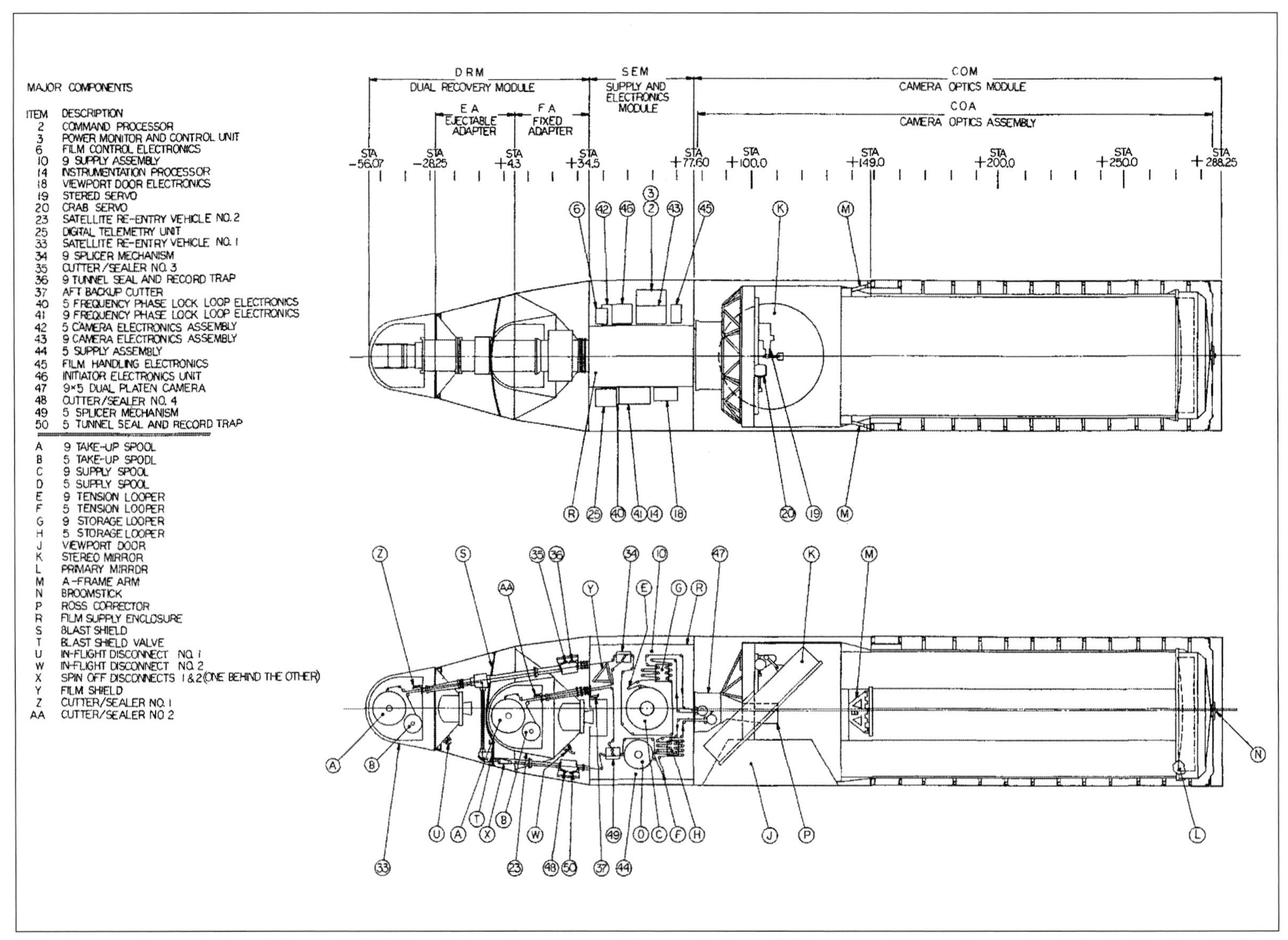

This diagram shows the breakdown of KH-7 Gambit imaging systems. They carried two SRVs, which delivered two rolls of film in 9-inch and 5-inch sizes back to Earth on a pair of spools to be processed and examined. The open view-port door (*J*) allowed light to bounce off the adjustable mirror (*K*) back to a primary mirror (*L*) of 44.5-inch diameter that weighed 347 pounds. It took ninety days to grind it to specs and another seventy days to polish. The light then bounced forward into the camera and corrective optics to expose both film rolls simultaneously. Film was then advanced forward onto the SRV spools, which would be cut when the spools were filled. The SRV was sealed and separated from the satellite to begin its journey home to an awaiting aircraft, to be snatched in midair. KH-7 is also known as the Gambit-1 series, used to follow up with high-resolution images after Corona passes that had photographed items of interest. *National Reconnaissance Office (NRO) Archive*

KH-7 Gambit open view-port door reveals the adjustable mirror taking up the left half of the interior scene, with the opening to the right down the camera optics module (COM) section to the primary mirror. The door would close to regulate thermal heating, and at times when the sun could shine into the imaging section and potentially damage components. Silver insulation is seen held in place with springs to keep it from obstructing the light pathway. Before the satellite orbital injection phase occurs, an external cover over the door is jettisoned. To the left of the mirror is the supply and electrical module, fixed and ejectable adapters, and the SRVs. *Air Force Museum, October 2016, by author*

KH-7 SRV with thrust cone containing the retrorocket. Also seen are the film pathways and electrical connectors that are severed once the SRV departs. White section to right is the fixed adapter of mission components. *Author*

A look inside the fixed adapter section contains hollow film guides, mission recorders, radio transmitters, a separation controller, and numerous wire harnesses. Open access at upper left is an orientation scanner. *Author*

Thrust cone has spin/despin thrusters that spin up the SRV to provide stabilization during maneuvers. The despin phase is used to stop spins during final preparations for reentry. A black thermal cover protects the SRV parachute. *Author*

KH-7 aft end shows (*at center*) a pair of thrusters, and an infrared band scanner (contained within round black housing) that senses the earth's horizon. It, along with the other scanners, provides proper alignment for taking photos. *Author*

KH-8 Gambit-3 was an improved KH-7. The forward section in top photo was connected to the aft Agena section by a rolling joint. This allowed the camera optics module to rotate and provided increased stability. Thinner film increased capacity to 12,241 feet. Resolution of photographed objects less than 2 feet wide was possible from altitudes of 65 to 90 miles. Life span on average was thirty-one days. Fifty-four missions were flown from 1966 to 1984. Early missions had only a single SRV, increased to two on later flights. Paint scheme was designed to provide even heating/cooling while in space. *Air Force Museum, October 2016, by author*

KH-8 aft section showing Agena engine, thrusters, and electronic-components arrangement. Not attached are the solar panels, providing electricity to run all the systems. Photo at lower left shows Model 8096 rocket engine, and large silver hydrazine propellant tanks with thrusters aft of these. Black boxes contain instrumentation operating several systems. Below right image taken at the forward end of the Agena aft section shows a pair of horizon sensors. Another pair is on the opposite side of the spacecraft body. *All photos by author*

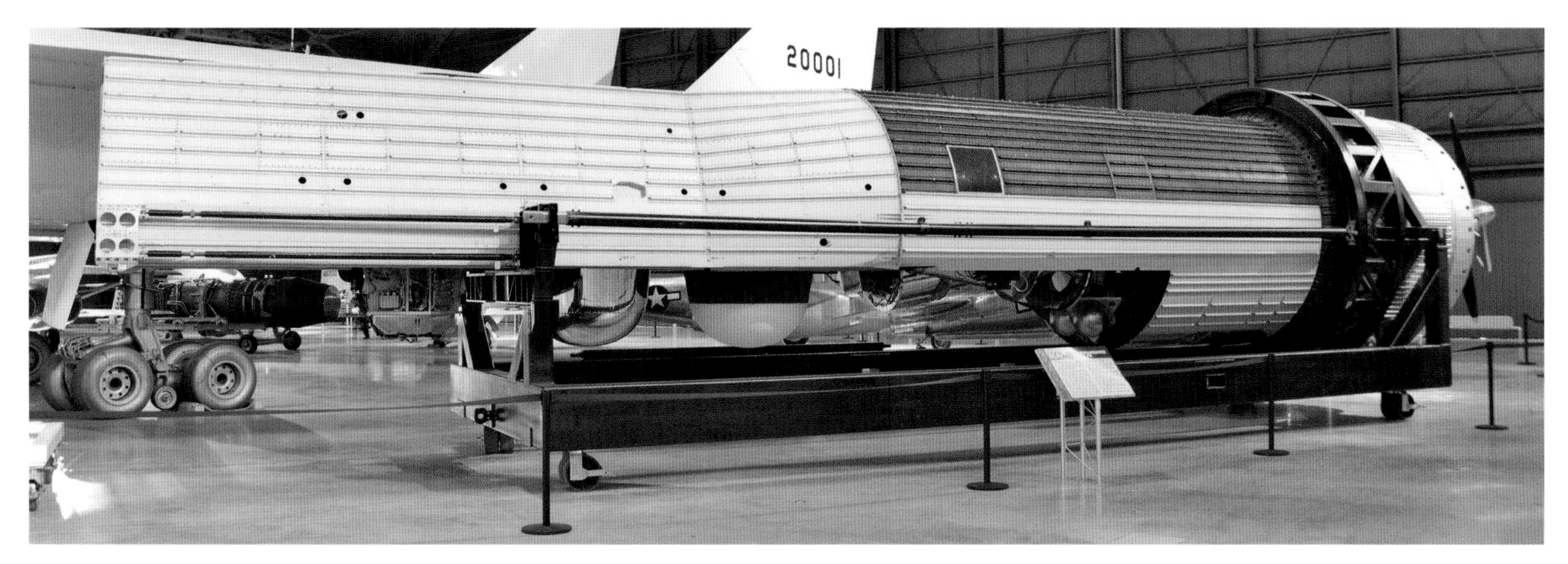

KH-9 Hexagon was 60 feet long and 10 feet wide and weighed 30,000 pounds. It was the last of the film return types used. Also known as "Big Bird," it carried four SRVs, plus a fifth in the mapping-camera module (not shown), attached to the forward end at the left edge of photo. Aft of SRVs are a pair of stereo cameras, of a design known as OBC (optical bar camera). The cameras shot wide-area imagery from 80 to 100 miles high, with detailed follow-up passes of objects of interest made by Gambit craft. KH-9s flew nineteen missions with an average life expectancy of 124 days. On board was 320,000 feet (60 miles' worth) of film. Each SRV could load 500 pounds of film (52,000 to 77,500 feet) before detaching for reentry and film processing. Bottom photos show, *to left*, a Mk. 8 SRV capsule inner section with film spools visible. Photo to right shows an SRV attached to the main vehicle; capsule container removed to show arrangement as well as film loaded to capacity. *All images by author*

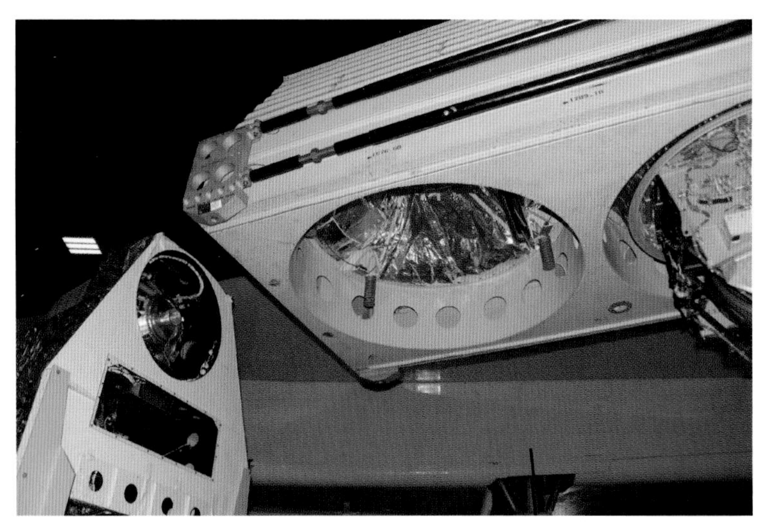

KH-9 forward SRV bay is empty. Springs visible are the means by which the capsule is ejected smoothly and evenly from the satellite. Silver insulation is used to regulate heat and cold extremes. *Author*

The SRV shown here is fully enclosed within the forebody ablative shield. As it heats and chars in the atmosphere, layers peel away, exposing new surfaces to absorb friction until the reentry phase ends. *Author*

Aft end shows small thruster quadrants as well as the larger center maneuvering thruster. Missing here are the solar panels that would be located in this section. Striped paint scheme is for thermal control. *Author*

Mapping camera would be attached to the end of the satellite seen in the background. It imaged areas of Earth so that a bomber could chart the safest corridors in and out of high-value targets. SRV brought this film back. *Author*

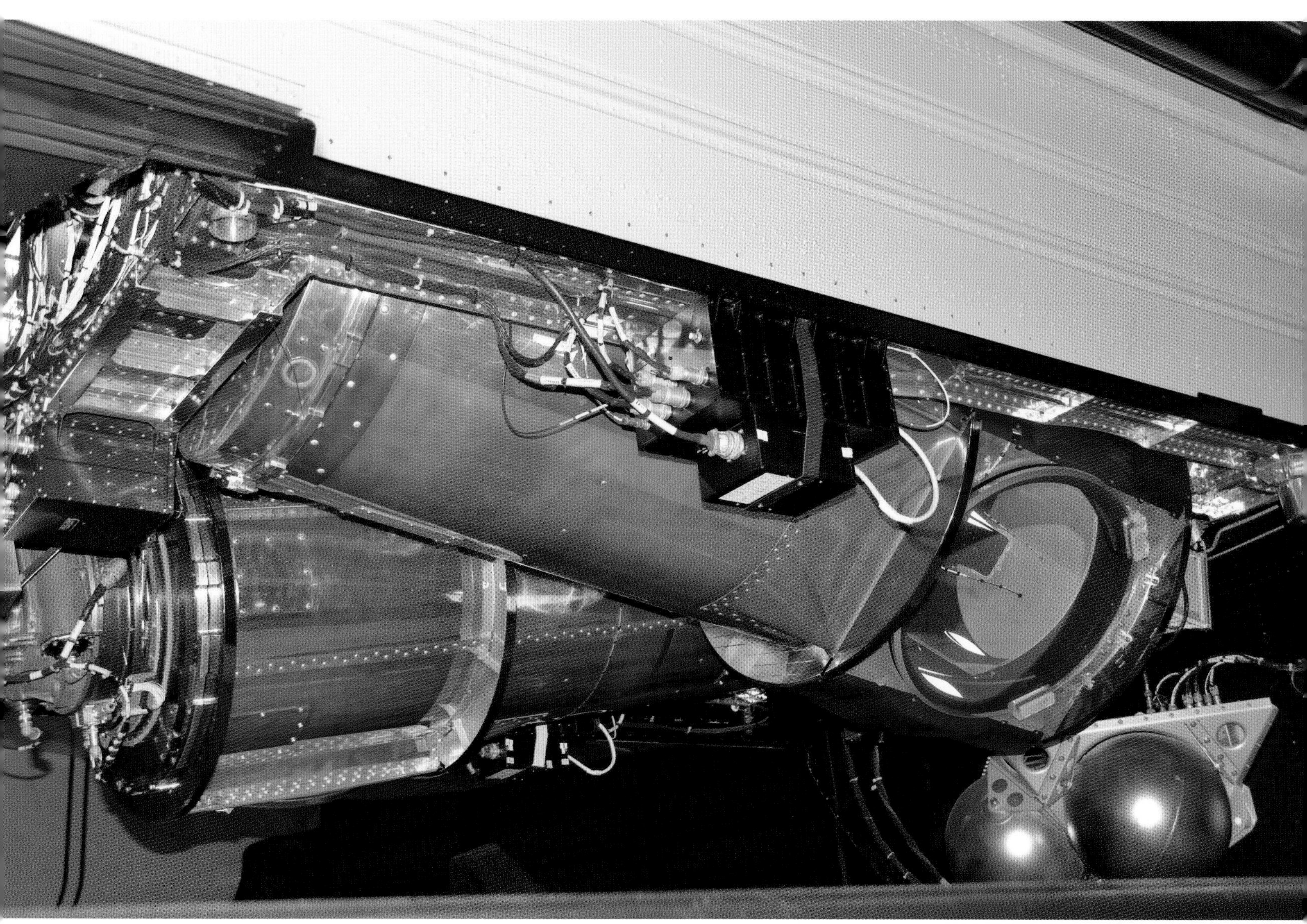

KH-9 HEXAGON A and B cameras in central bay section. They worked together, shooting in stereo mode, the film being advanced with each exposure. The drive mechanisms are seen at left. Cameras are angled so that part of their fields of view overlap slightly. Light entered the round, optically clear window seen at right end of foreground camera, which also corrected for any spherical aberration present. The mirror behind the window angled the image to a 36-inch concave main mirror, which reflected the image to the film-exposing components. *Author*

Two views inside the missile gallery of the USAF Museum, in Dayton, Ohio, showing the DSP (defense support program) satellite structural test article. These full-scale items are used by engineers and designers to fit and check out various components that make up entire satellites before they are launched and operating. Now the standard platform, replacing the earlier MIDAS and VELA series, twenty-three have been observing Earth since November 1970. The large telescope seen in the photos is offset at an angle of 7.5 degrees from the central axis as the satellite spins slowly at 5.7 rpm in geostationary orbit. This allows the infrared sensors (6,000 of them in the latest production series) to cover the entire earth, reporting any irregularities to ground stations in 1.2 seconds. The platforms are hardened against the EMP effect and have protective filters to prevent blinding by hostile lasers or particle beams. *Left photo, October 2016; right photo, October 2018 by the author*

Eventually, antisatellite (ASAT) weapons were developed, tested, and improved on over time. The ASM-135 could be launched by F-15 Eagle fighters in a direct-ascent launch profile. It was successful on September 13, 1985, when the Solwind P78-1 satellite was destroyed orbiting at 345 miles in altitude. The aft section was a converted SRAM (short-range attack missile) booster. Length was 17 feet, 9.5 inches; weight was 2,600 pounds; and speed was 15,000 miles per hour. These two are of the CASM-135, a captive-carry inert-training shape. The top one is at the Hazy Center; the bottom one is shown at the USAF Museum. They were banned from deployment due to costs and treaty implications. *Author*

The Homing Overlay Experiment (HOE) was a US Army program designed with the hit-to-kill method of intercepting ICBM reentry vehicles without using a conventional or atomic warhead. A kinetic impact would damage or destroy targets. This example seen at the Hazy Center is 17 feet long and 3 feet in diameter and weighs 2,400 pounds. After launch and lock-on by a suite of long-wave infrared sensors of the target, a radial net composed of thirty-six aluminum ribs unfurls, which increases the probability of a hit. On June 10, 1984, a Minuteman launched a simulated reentry vehicle (RV) into space. The HOE was fired from Kwajelein Atoll in the Pacific Ocean. A successful intercept was made at a closing speed of 20,000 feet per second at an altitude of over 100 miles. *July 2013 by the author*

Two miniature seeker heads of early designs are shown, both displayed for public viewing at the Hazy Center. On top is the Exo-atmospheric Kill Vehicle (EKV). It was the payload of ground-based interceptor (GBI) missiles. After launch and separation from the rocket, the EKV would search for and lock onto a target. It was the same objective for the device in the bottom image, the lightweight exo-atmospheric projectile (LEAP). Both had no warheads, only the mass of the seeker, which was used to physically smash into satellites or hostile warheads. This alone can cause enough damage to prevent the targets from successfully continuing to function normally. LEAP was designed for use on Standard SM-3 missiles, deployed at sea on warships of the Ticonderoga and Arleigh Burke classes. Satellites and missile warheads are not especially designed to withstand impacts, since such shielding would add weight, reducing rocket range capability or "throw weight" performance. Advances in technology have enabled these seekers to be of minimal weight, capable of making midcourse corrections in microseconds as they move at thousands of miles an hour toward their targets. Thruster arrays seen on both allow for such rapid course corrections. They can be fitted to a wide assortment of missiles currently in use, and their quick-reaction capability does not give intended targets time to maneuver or change orbit to evade them. Despite earlier bans on their use, they are now back in frontline inventories due to rogue nations having the ability to launch their own satellites. Some have developed and tested missiles of short and intermediate ranges with nuclear warheads, destabilizing the long-established Cold War–era superpower balance. They also have another purpose: they can destroy malfunctioning satellites that have hazardous materials aboard, or top-secret equipment not intended to be examined by hostile nations if it were to reenter the atmosphere and crash on land. An example of this actually occurred when the spy satellite USA 193, launched in December 2006, ended up out of control and in a low orbit; it contained 999 pounds of hydrazine, a dangerous rocket fuel. It was destroyed by USS *Lake Erie* in February 2007, equipped with SM-3 missiles containing interceptor warheads. *Photos by author*

Chapter 4

PROJECT APOLLO

Project Gemini proved that humans could exist in space for the time estimated to go to the moon and back, accomplishing tasks while staying healthy all the while. Ranger photos, Surveyor soft lander data, and Lunar Orbiter scouting of potential landing sites all set the stage for the Apollo Project to send men to the moon and bring them home. This was not simply an earth-orbit adventure. Reaching orbit was but a single step, since there was the outbound journey, a braking event behind the moon, docking/undocking events, the landing and exploration phases, takeoff, and the return to Earth. Apollo 7 was the shakedown flight after the Apollo 1 tragedy. Apollo 8 went to the moon to orbit and return. Apollo 9 checked out the major components in earth orbit. Apollo 10 was a full-dress rehearsal, with the lunar module coming to within 10 miles of a landing. Apollo 11 was the ultimate voyage.

Thankfully, many people with a vision of the future have stored, preserved, and displayed many of the artifacts from our moon efforts, which can be seen today in many facilities across the nation. Photographs of the wide variety of relics from that episode are proof that much can be seen and learned as a result.

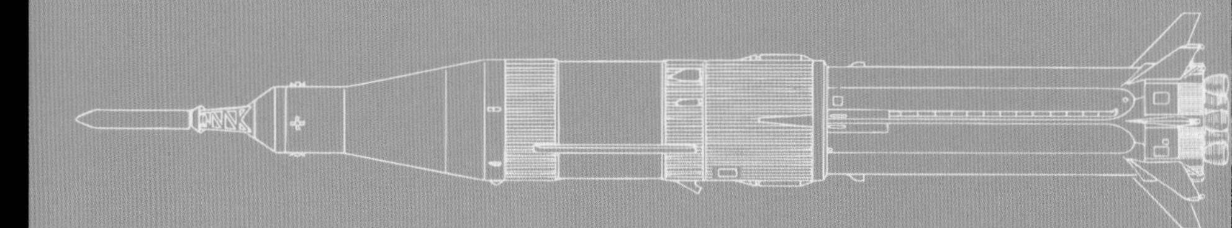

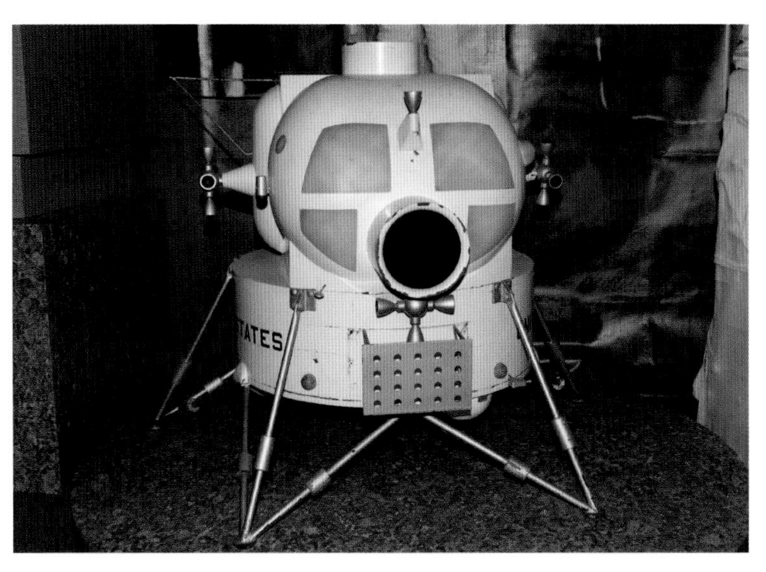

As with any monumental project, conceptual models are a necessity to show others what the current state of technology allows for, to inspire thought and ideas on what is missing or is worth further study. The left picture shows an Apollo command module; this carried the three astronauts to the moon and back. It is basically a scaled-up Gemini with all the extras needed to sustain the crew plus allow them to navigate and control the craft. It had to dock with the lunar module and was the only part that came back to Earth. The right image shows a 1962 vision for what the Lunar Excursion Module (LEM) could be like, with a descent section and ascent stage where two astronauts would ride down to the surface. Thrusters, hatches, and window arrangement would all change quite a bit over the years. Models seen above, photographed by the author in January 2019, are of early 1960s vintage and are displayed at the American Space Museum & Walk of Fame in Titusville, Florida. The bottom illustrations are NASA prints of the three manned spacecraft series.

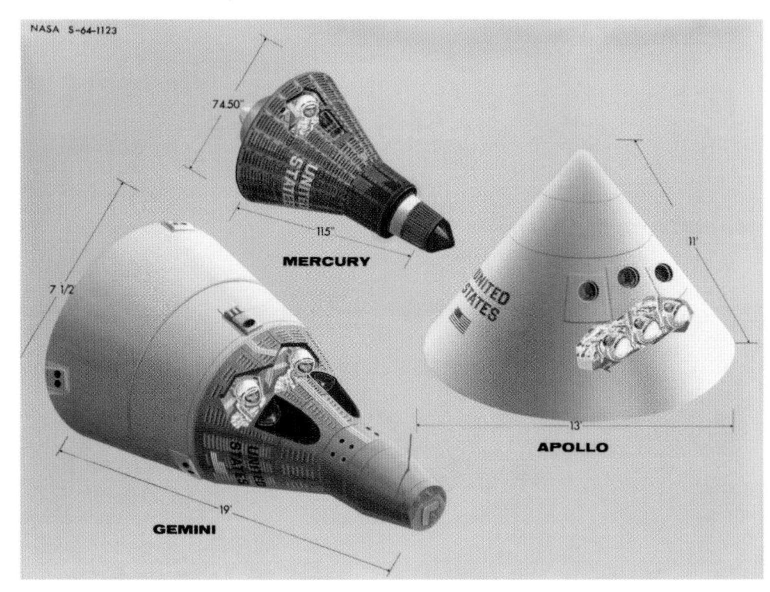

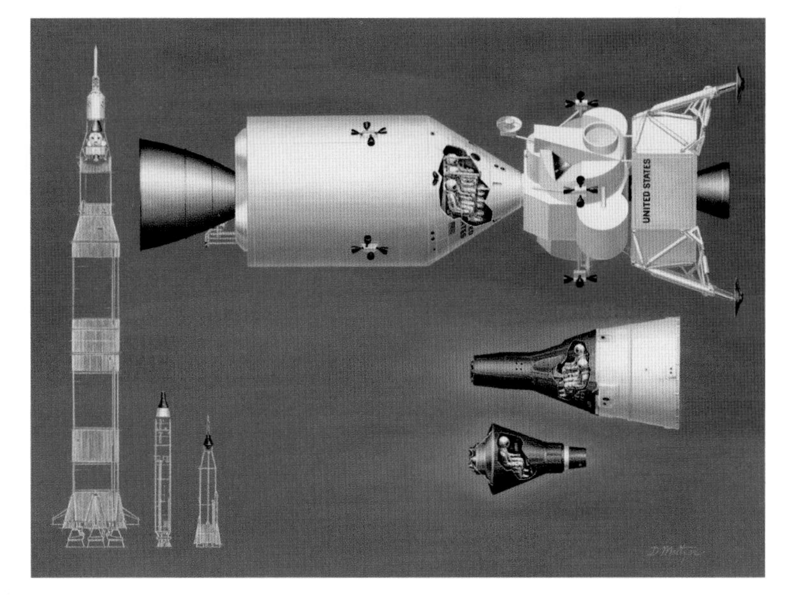

The VAB (vehicle assembly building) with the LCC (launch control center) in foreground. Both were built in the 1960s to put together and launch the Saturn V rockets. *Author*

The crawler way at Kennedy Space Center from the VAB to the launchpads. River stones line the surface the crawler moves on while transporting spacecraft, since the ground is too soft without adequate reinforcement. *Author*

Kennedy Space Center (KSC) bus tours often reveal activity on the site such as a mobile launch platform (MLP) modified for space shuttle use on top of a crawler. There are two crawler transporter (CT) vehicles. *Author*

An MLP at rest in the staging area. Crawler is driven under it and secured, then is free to be positioned around the launch complex from inside the VAB to a trip to one of the launchpads. *Author*

Not all Saturn rocket launches were of the moon-capable vehicle. Early missions had a first stage of eight outer fuel tanks adapted from Redstone rockets. The inner tank was manufactured from Jupiter rocket tooling. At left is the Project Highwater Saturn-Apollo 3 launch from Complex 34 (November 16, 1962), which released 23,000 gallons of water at an altitude of 104 miles. At right is the Saturn-Apollo 6 launch on May 28, 1964, from launchpad 37B, the first liftoff of a boilerplate Apollo capsule. It orbited the earth four times. The follow-on Saturn IB used an uprated first stage with an S-IVB second stage. *Both photos, NASA*

PROJECT APOLLO: MANNED MISSIONS

Date	Mission	Crew	Duration
October 11. 1968	Apollo 7	Schirra/Eisele/Cunningham	11-day earth-orbit-proving flight
December 21, 1968	Apollo 8	Borman/Lovell/Anders	Saturn V test flight to moon and back
March 3, 1969	Apollo 9	Scott/McDivitt/Schweickart	LM shakedown test in earth orbit
May 18, 6199	Apollo 10	Young/Stafford/Cernan	moon-landing dress rehearsal
July 16, 1969	Apollo 11	Armstrong/Aldrin/Collins	first moon landing, Sea of Tranquility
November 14, 1969	Apollo 12	Conrad/Bean/Gordon	moon, Ocean of Storms
April 11, 1970	Apollo 13	Lovell/Haise/Swigert	mission aborted; crew safely returned
January 31, 1971	Apollo 14	Shepard/Mitchell/Roosa	moon, Fra Mauro
July 26, 1971	Apollo 15	Scott/Irwin/Worden	moon, first J mission with LRV, Hadley Rille
April 16, 1972	Apollo 16	Young/Duke/Mattingly	moon, Descartes/Cayley Plain
December 7, 1972	Apollo 17	Cernan/Schmidt/Evans	moon, Sea of Serenity / Taurus Littrow

Apollo program total cost: $25.4 billion from 1961 to 1973. Adjusted for inflation in 2020 dollars, the cost would have been $194 billion.

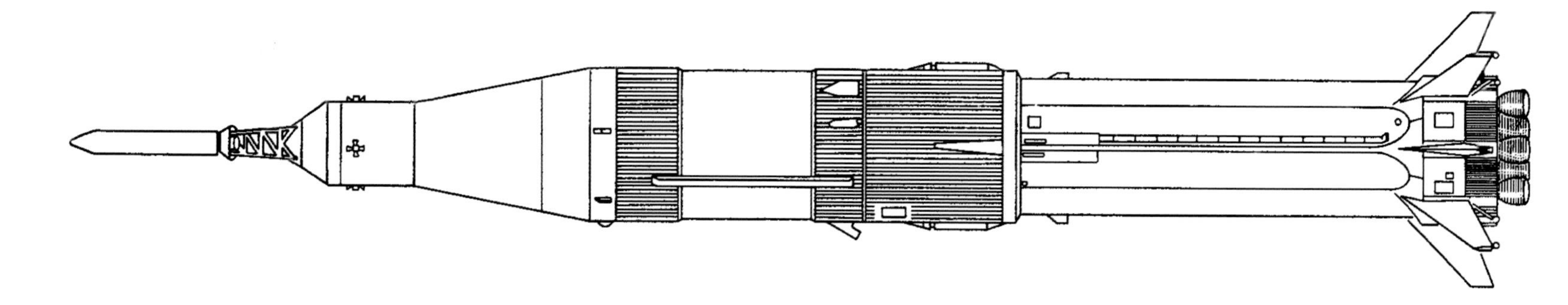

Saturn IB NASA line drawing. Manned flights with this variant of the Saturn series totaled five.

At the Kennedy Space Center rocket park is this Saturn IB SA-209, shown horizontally in July 2019. This variant had the uprated first stage and the S-IVB upper stage. Lower left photo shows details of the launch escape tower (LES), the command module (CM) and the service module (SM) including the quad arrangement of maneuvering thrusters. At upper right is a gray APS (auxiliary propulsion system) pod used for stabilization of the stage during lunar module extraction. At the bottom of the first stage are eight H-1 engines and fins. This version flew nine times. Four proving flights ahead of the moon launches, Apollo 7: three Skylab flights and one for the Apollo-Soyuz Test Project. One was kept in readiness in case of a Skylab rescue mission but was never needed. *Author*

The Space & Rocket Center in Huntsville, Alabama, has this early-configuration Saturn I Block I example outside in a vertical assembly. A Juno II rocket is to its right. It flew ten times, with the ability to lift 20,000 pounds into low earth orbit. On US interstate highway I-65, close to the Alabama-Tennessee border, is this Saturn IB at a roadside rest area. It is accessible only from the southbound direction. *Both pictures were taken in October 2016 by the author*

A Saturn IB H-1 engine, of which eight were used in the first stage. At photo center is the tank containing turbopump lubricant. Gold-colored, four-bladed impeller is the liquid oxygen (LOX) inlet, at the center of the LOX pump. Silver ribbed pipe below it is the LOX feed line. Black assembly at end of engine is the gimbal bearing. Length is 8.8 feet, with a 4.9-foot diameter. Weight is 2,200 pounds. Propellants burned were LOX oxidizer and RP-1 as the fuel. Thrust at sea level was 200,000 pounds. *Huntsville, October 2016, by the author*

The ultimate Saturn variant was the Saturn V moon rocket. This vertical example is at the Space & Rocket Center, which is a replica. This same facility also has one stacked horizontally indoors, along with ones in Houston and at KSC in Florida. Those three should have gone on moon missions as Apollo 18, 19, and 20 but were canceled due to budget conflicts in the early 1970s. The example in Houston has flight-rated hardware, with the other two composed of many test pieces and mockup components. Despite this, all three locations have done excellent jobs preserving and arranging the rockets for the public to admire and learn about. The Saturn V stands 363 feet tall, and fully loaded it weighs 6.2 million pounds. *Author photos, October 2016*

The horizontal Saturn V in Alabama has at the forward end the launch escape tower with command module attached. Capsule markings were used during tests of the escape system. *Author*

Command Service Module (CSM) 119 at KSC, intended for use as the Skylab rescue flight. The capsule was modified for a crew of two (Vance Brand and Don Lind), returning with the three additional crew. Rocket would have been the SA-209 Saturn IB. *Author*

At KSC, the CM/SM interface is shown. Also visible are the domed ends of the propellant tanks. A truss fairing providing connections for water and electricity is not seen here but is visible on the bottom of the capsule in top right image. *Author*

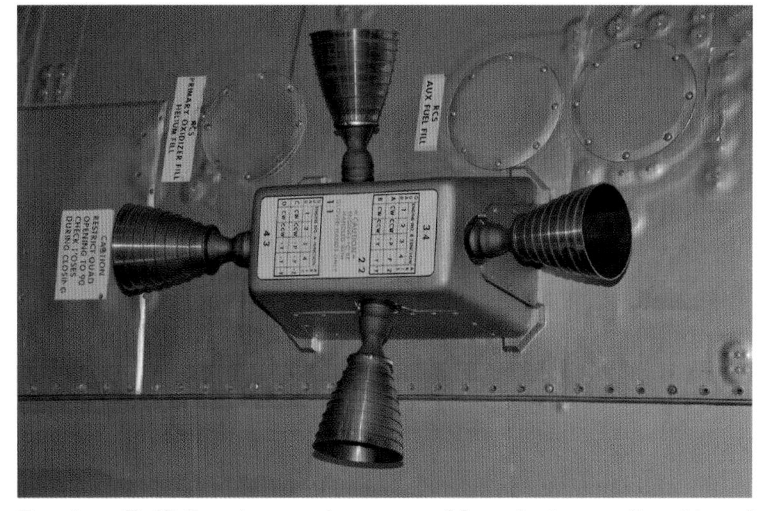

Four type R-4D thrusters made up one of four clusters on the sides of the service module, used for maneuvers and backing away from the S-IVB third stage after docking with the lunar module. Each one provided 100 pounds of thrust. *Author*

The Hazy Center has exhibited the upper half of the service module SPS (service propulsion system) engine. This engine played critical roles in braking the docked modules as they passed behind the far side of the moon to enter orbit, and for escaping the moon's gravity to begin the TEI (trans-earth injection) maneuver to head back home. It used a fuel of aerozine 50 (unsymmetrical dimethyl hydrazine [UDMH]) and nitrogen tetroxide as the oxidizer. No fuel pump was used; helium gas provided system pressure. Engine thrust was rated at 20,500 pounds. At KSC, the lower half of the SPS engine was the large exhaust nozzle. On the CSM-119 example, the original data plate is still riveted to the outside edge of the nozzle. *Both photos by the author*

Before the Saturn V rocket flew, scale wind tunnel models were built and "flown" to test phases of flight within Earth's atmosphere. The top model was used to investigate escape tower exhaust plumes on the command module structure. Both items were heavily instrumented with sensors and gauges, connected to diagnostic equipment so values and stresses could be recorded and evaluated after a test to ensure predicted values aligned with actual test results. *Top photo, Hazy Center, July 2013; bottom image, Space & Rocket Center, by author*

The Kennedy Space Center Apollo exhibit shows an SLA (spacecraft lunar module adapter) section that contained the lunar module. The top photo partially shows the silver service module body (*to left*), with the top section of the S-IVB third stage to right. Black-striped section is where the instrument unit (IU) was located within. The adapter section had four panels that separated once the CM/SM separated from it, to expose the LM after the translunar injection (TLI) maneuver was completed. The command/service module would then turn 180 degrees, line up on the LM, and move in to dock. It would then pull the LM free of the spent stage and continue on to the moon. The bottom picture shows a scale Saturn V model, with clear panels showing the LM enclosed with legs folded, and the service module SPS engine nozzle directly above it. Large red tank is part of the third-stage rocket fuel supply. Many concepts of transporting these vehicles was discussed and calculated, this method being the most efficient and achievable. *Author*

One can see the instrument unit (3 feet high, 22 feet in diameter) location clearly at Huntsville between the SLA and third stage. This section contained the brains of the Saturn V rocket. It flew the vehicle on a programmed trajectory out of the atmosphere into orbit and controlled staging, among many other things. Labeled items and other electronic components were mounted to cold plates (brown panels) that absorbed heat. The external power connection was an umbilical cable housing that disconnected at liftoff. *Author photo, October 2016*

Instrument unit ST-124M inertial platform, as mounted in place. ST indicated "stable table," with "M" for moon. Silver sphere at right was a nitrogen supply source for the platform frictionless bearings. *Author*

The FCC with wire bundles connected. It controlled the gimbal adjustments made to the engines. It received data from the LVDA unit, translating that into corrective steering commands to each engine. *Author*

An interior view of the ST-124M shows three gyros. They measured changes in the X, Y, or Z axis, with any deviations corrected for to keep the rocket on course. Exotic metals such as beryllium and elkonite were used inside. *Author*

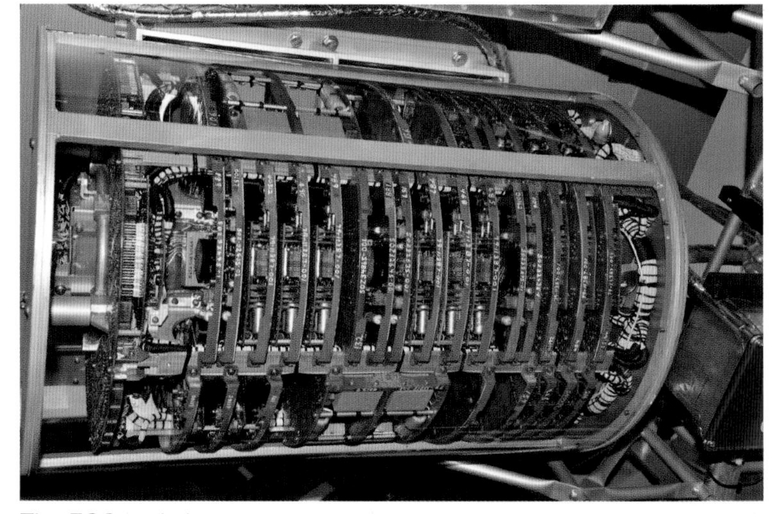

The FCC had circuit boards stacked inside it. As an analog device, it received a continuous LVDA electrical signal of variable amplitude. Unit was sealed with provisions made to regulate heat accumulation. *Author*

The S-II second-stage number 14 at KSC (*in the top photo*). It would have flown on Apollo 19 if that mission had not been canceled. This and the other Saturn rocket artifacts, which are inside at present, were outside for years, close to the VAB. To avoid expensive refurbishments, especially being on the Atlantic coastline, with its corrosive salt air, it was decided to place all the items inside, which was the best option for long-term preservation. Length is 81 feet, 7 inches, and diameter is 33 feet; stage weighs 95,000 pounds dry / 1,037,000 pounds loaded. Each of the five J-2 engines produces 225,000 pounds of thrust. Four outer engines are gimbaled; inner one is fixed in position. Fuel capacities are 260,000 gallons of liquid hydrogen and 830,000 pounds of liquid oxygen. Stage burns continuously for six minutes. An interstage ring (not shown) surrounds the engines and attaches to the top of the first stage. This falls away after eight ullage engines fire to move the stack ahead of the spent first stage. The bottom picture shows the S-IVB third-stage aft end with its single J-2 engine. It was gimbaled and could be restarted in a vacuum. It burned liquid oxygen and liquid hydrogen at 230,000 pounds of thrust in space. After the LM was removed from the upper section of the stages, they were sent into orbits around the sun. Some impacted the moon to check seismometers left by Apollo missions. Gold-colored cone fitting over the aft liquid oxygen fuel tank dome is a thrust structure designed to distribute the thrust loads evenly when the engine is firing. Entire section seen here is enclosed within an interstage fairing connecting to the top of the S-II second stage. *Top picture, KSC, November 2011; bottom picture, March 2017, by the author*

A second/third-stage interstage fairing at the Space & Rocket Center has been modified for use on Earth. Visible is the construction and two of the black forward-firing retrorocket fairings to slow second stage down after burnout. *Author*

Some of the eight first/second-stage interstage ullage (actually, they are solid-fuel) rockets: 12.5 inches in diameter, 89 inches long, 12,500 pounds thrust for four seconds at first-stage burnout. This keeps the stack moving forward plus moves fuel aft into the J-2 engines. *Author*

The Saturn V model at KSC shows the red second/third-stage interstage retrorockets, which decelerate the spent stage away from the third stage. This is an electrical command once the second-stage engines have used up their fuel. *Author*

Space & Rocket Center Saturn V model shows an outer first-stage fairing that contains red retrorockets. They fire to slow the empty first stage down, providing a safe separation phase from the interstage before the second stage ignites. *Author*

This is the view visitors to the KSC Apollo exhibit see upon entering the building, which is the aft end of the first stage. Five F-1 engines were key elements in the Saturn V design. Without them, the rocket would not have been able to get off the ground. The stage is painted to represent the one used on Apollo 11, but in reality that stage is at the bottom of the Atlantic Ocean. This is the S-1C-T stage, used to test the design. A clue is the two gray pipes visible at the nine and three o'clock positions on the white heat shield in the left photo, used to feed fuel into the stage during live firings. These pipes are not on a flight-rated stage. Each F-1 engine produced 1.5 million pounds of thrust at liftoff; one engine weighs 18,416 pounds and is 18.5 feet high. At liftoff, each burns 3 tons of fuel each second. In right image, ribbed line that wraps around the engine nozzle is the turbine exhaust. A white gimbal actuator is also seen. Bottom left photo shows red fuel and oxidizer feed lines. Bottom right picture shows thrust chamber. Tubes lining the inner wall are coolant paths to prevent nozzle overheating. *Bottom photos, Space & Rocket Center*, *all by author*

The massive S-1C first stage (this one seen at the Space & Rocket Center) weighs 288,800 pounds dry / 5,030,500 pounds full. Diameter, including fins at base, is 63 feet; length is 138 feet; liquid oxygen capacity is 346,372 gallons. RP-1 kerosene requirement was for 212,846 gallons; total weight of fuel is 4,733,924 pounds. First-stage continuous duration of operation from liftoff is 167.3 seconds. Large square panel in white paint just forward of the red USA marking is an access for intertank maintenance. *Author*

Command modules are displayed across the nation. This is the most famous of them all, *Columbia*, of Apollo 11, at the National Air & Space Museum. It is fully enclosed inside Plexiglas. *Author*

Apollo 15 CM *Endeavour* is at the National Museum of the United States Air Force in Dayton. It was the first of the Apollo-J series, modified for extended days in lunar orbit, with a roving vehicle, cameras, and sub-satellite in the SM. *Author*

Apollo 14 *Kitty Hawk* at the Kennedy Space Center. Handholds are for EVA assistance. Window right of open hatch was used primarily for docking alignment with the lunar module *Antares*. *Author*

Apollo boilerplate (non-flight-rated) CM at the Hazy Center shows components of the Apollo 11 post-splashdown configuration after US Navy divers attached the stabilization collar. Also seen are the Apollo 11 flotation bags used to right an upside-down CM in the water. *Author*

Apollo 16 CM detail shows empty mortar canisters for the drogue chutes. One of the main chutes would be in the void space to the left of the canister. The pair of round openings close together (*lower right*) are pitch thrusters. *Author*

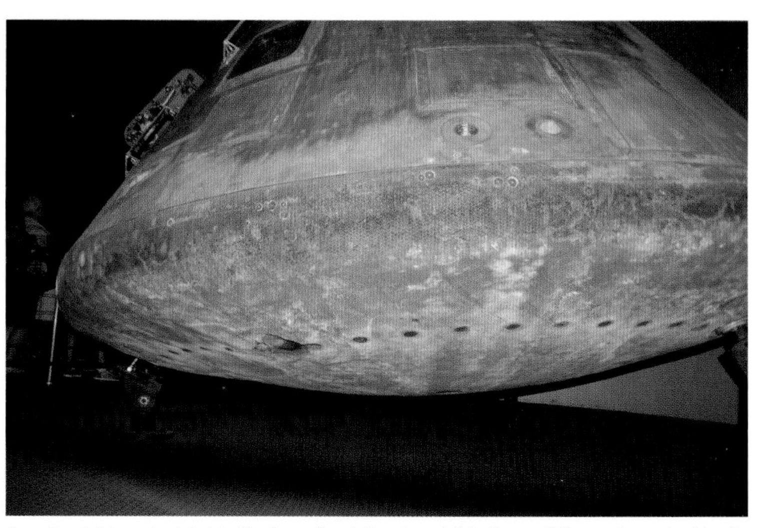

Apollo 14 heat shield. It absorbed the heat friction of the return to Earth as it ablated or flaked off layers of material to keep the interior cool and the spacecraft intact. These capsules weighed about 12,000 pounds on average. *Author*

One of the main parachutes is seen on the Apollo 14 CM. This is an example of how a predeployed one would have appeared before use. There were three main chutes total to lower the capsule to splashdown. *Author*

To navigate in space between the earth and moon, a telescope and sextant were used. These apertures are seen on the Apollo 15 command module. Location was the opposite side of the entry hatch. *Author*

The Apollo guidance and navigation (G&N) device, displayed at the Air & Space Museum. Eyepieces for the sextant and scanning telescope are seen, with the display keyboard (DSKY) just right of center. *Author*

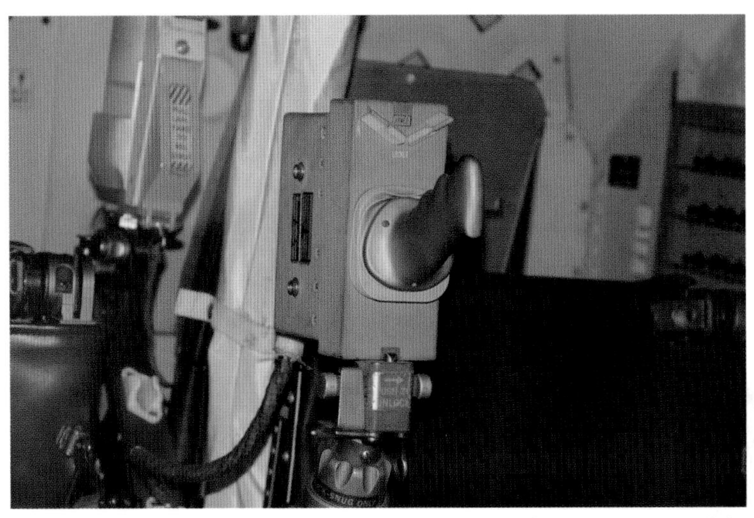

Apollo 14 CM rotational hand controller. Using it fired service module thrusters to adjust CM/SM orientation. There were two controls, one at the right (LM pilot) seat and one at the CM pilot position. *Author*

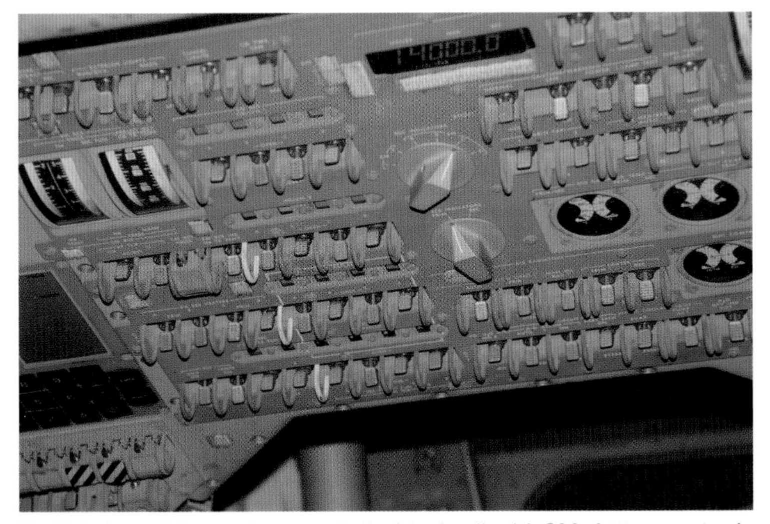

Partial view of the main console inside Apollo 14 CM. At top center is the mission timer. Two round dials select environmental and reaction control gauges. All switches have guards to avoid accidental use. *Author*

Apollo 14 view of CM pilot seat at center, with the LM crewman station to right. Triangular panel at upper right is for environmental controls. The spaces under the seats contain numerous electronic components. *Author*

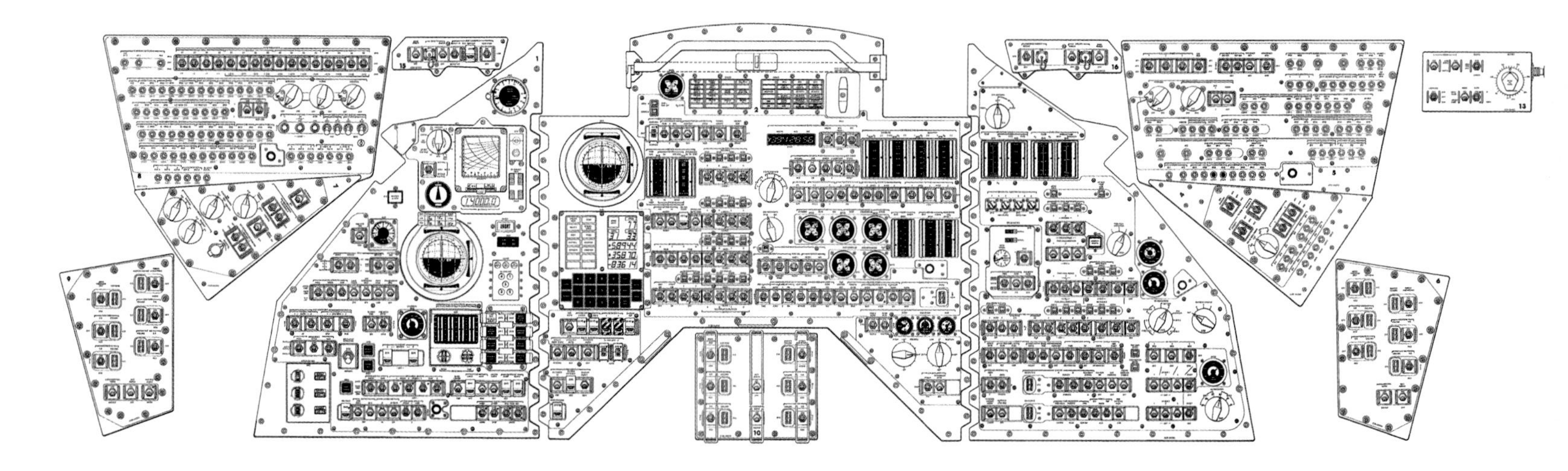

The main instrument panels of Apollo Command Modules 112 through 114 (Apollo 15/16/17) are shown in the central line drawing, with side panels on their appropriate ends of the illustration. Main panel is divided into mission commander flight controls on the left section; the central section in front of the command module pilot deals with caution and warning systems as well as reaction control. He also has the guidance and control display keyboard (DSKY) panels below one of the pair of round flight director attitude indicators. The top of the central console has the ingress/egress handle, with the warning annunciator panel lights below it. The right-hand panel addresses electrical power / fuel cells as well as the Service Module Propulsion System (SPS) functions. Large end panels are primarily circuit breaker locations. Layout is designed to be used with gloves and within easy reach of each crewman working his station while strapped in his seat. In total, the panels contain seventy-one lights, forty event indicators, twenty-four instruments, and 566 switches. To the left is a typical display keyboard, used by the crew to interact with the guidance and control systems. *Drawing, NASA; photo, Space & Rocket Center, by the author*

At the Kennedy Space Center visitor complex, in the Apollo-Saturn gallery is Lunar Module 9, suspended from the ceiling as if about to touch down on the moon. LM-9 was an H-series variant slated to fly on Apollo 15, but it was substituted for a J-series LM that could carry the lunar rover and remain on the moon for a longer time. This is only one of three LMs of flightworthy status on display. This example clearly shows the amber-colored aluminized kapton foil enclosing the decent stage. The ascent stage was mostly bare aluminum, of 2024 and 5056 alloys. Kapton tape was used on the struts of the landing-gear legs. The footpads are missing the lunar contact probes, which alerted the crew to shut down the descent engine as the craft was about to land. Large, round, white radar at very top is for rendezvous and docking. The steerable antenna for S-band reception/transmission is pointed to the left. The ascent stage egress/ingress hatch above the ladder is open. Large, round light above hatch is a flashing beacon used for sighting the ascent stage by the CM commander prior to docking. *November 2011 by author*

The lunar module was the product of ideas formulating in the early 1960s on which method would be used to land on the moon. The direct landing could be achieved by building a huge rocket informally called Nova, where there would be no docking maneuvers. The entire Apollo CM, SM, and LM all would go to the moon and land as a single unit. Another idea was to use two Saturn V rockets: one to launch the spacecraft, and another to orbit with a load of fuel the crew would top off from and then head for a lunar landing. In the lunar surface rendezvous scheme, an unmanned lander would carry everything, including fuel, to a landing site. Another manned vehicle would meet up with the lander, conduct EVAs, fill up with fuel, then blast off to head home. The other centered on a lunar orbit rendezvous (LOR) flight plan. In this scenario, the Saturn V would carry everything into earth orbit, then the CM/SM would dock with the LM and continue on to lunar orbit, with the LM making the landing. LOR was not considered a safe method, since all the maneuvers and dockings were risky and complicated. Eventually, ardent supporters of LOR and even Dr. Wernher von Braun himself became converts to the idea. The chart below shows the instrument panel layout for LM-6, dated October 1969, including the hand controllers for operating the engines and thrusters. Most of the needed systems were installed outside the crew cabin, making the vehicle large. With weight the overriding priority throughout development, a compromise was reached where the most-vital systems requiring crew access to diagnose and repair were made accessible. This resulted in a smaller ascent stage. The descent stage was fairly straightforward in layout and assembly, once engine designs and fuel requirements were frozen. In the end, the lunar module was another of the many innovations to come out of the Apollo program. There was no historical precedent to base the LM design on, since it was to operate strictly in space and in one-sixth gravity. *LM-6 control and display panels chart by NASA*

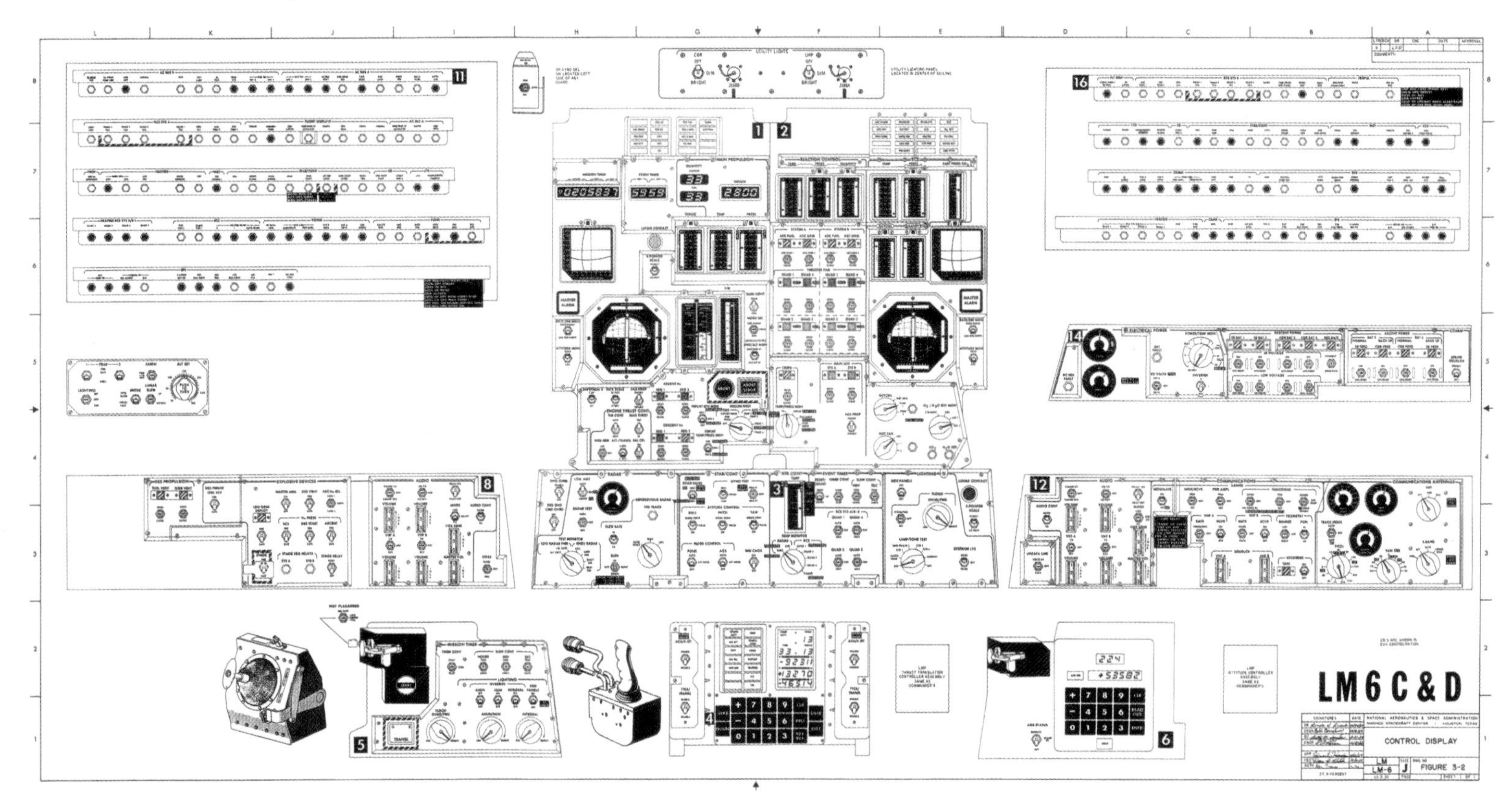

Lunar Module 2 is at the National Air & Space Museum and can be viewed from above. This angle of the ascent stage shows black panels used as a means to regulate temperature. *Author*

The rendezvous radar, with a VHF antenna behind it. Also, to the left of that is the docking target, used to visually align the command module as it moved in to connect the two craft together. *Author*

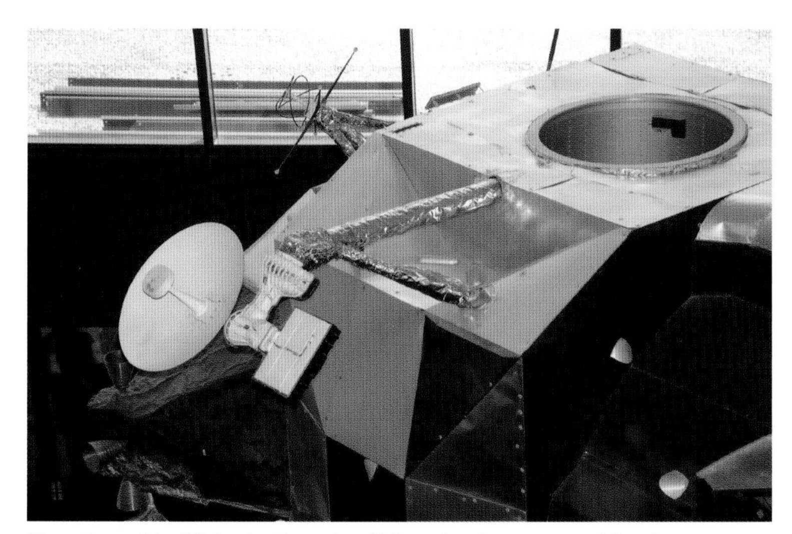

The steerable 26-inch-diameter S-band antenna, used for deep-space voice and telemetry links. It can be used in a range from 174 degrees in azimuth to 330 degrees in elevation. The docking tunnel is to the right. *Author*

One four-thruster group located at the right back or starboard aft corner of the ascent stage. Spiral-shaped antenna is one of two in-flight S-band units; another is forward. At upper left is a partial view of the deployed EVA antenna. *Author*

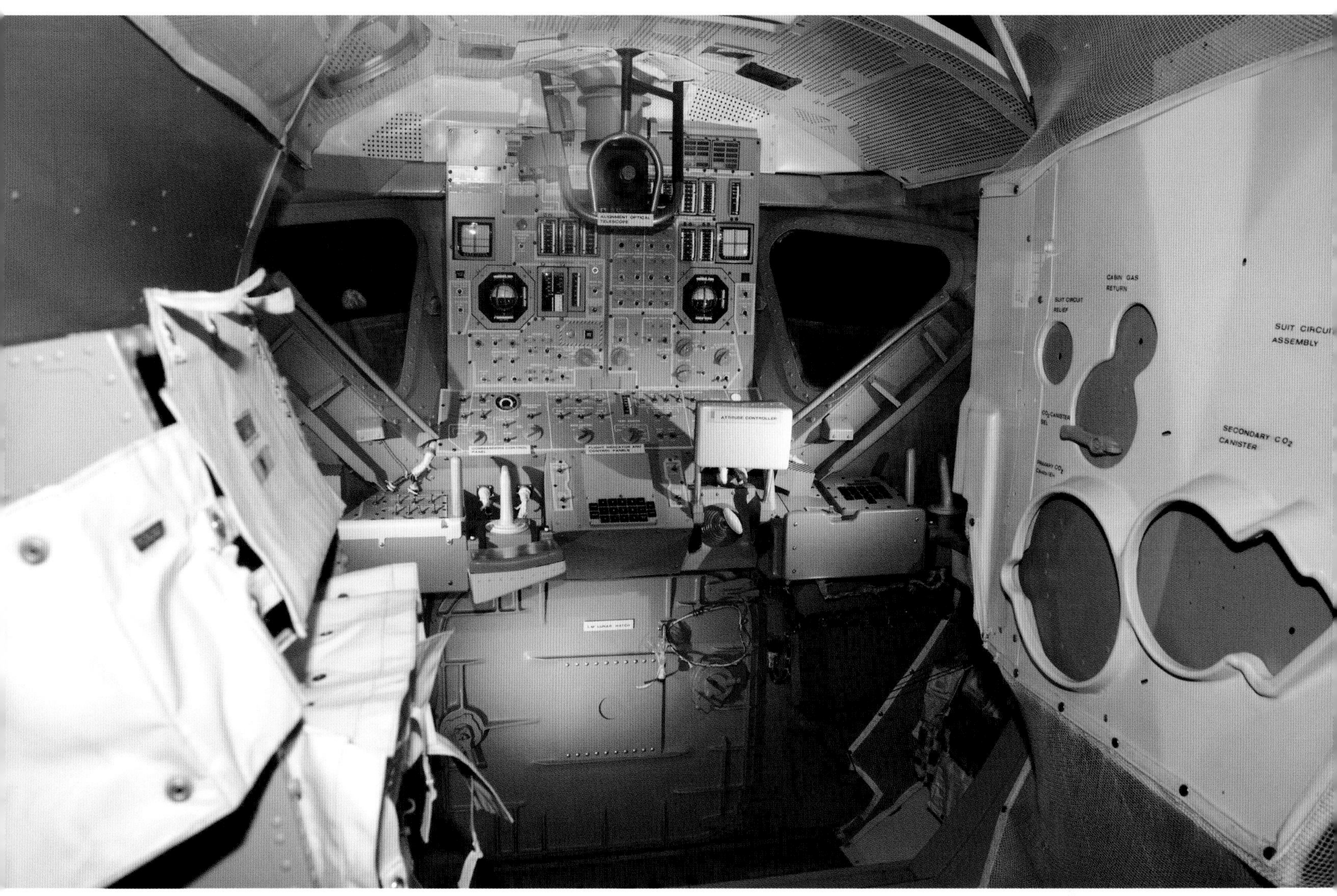

At the Kennedy Space Center, you can get a look inside an LM ascent-stage simulator interior and imagine how it was for the twelve men who used the LM on six occasions to land on the moon, live in, and leave the moon when their surface duties were completed. There were no seats; the two crew positions had bungee cords attached to their spacesuits to give them stability as they went through landing and takeoff events. The alignment optical telescope is the upper center device, with a guard that could also be used as a handle. The egress/ingress hatch at lower center was hinged to open to the right. The camera position for this picture is approximately where the ascent engine cover and aft cabin bulkhead were located. Left and right cabin sidewalls were used for equipment storage. The livable space amounted to 160 cubic feet. *Author*

In front of the egress hatch was the “porch” section. It was used to guide the astronauts to the ladder below it. Yellow handle was pulled to release the MESA—Modular Equipment Stowage Assembly—in quad bay 4. *Author*

LM-2-landing radar unit. It could be adjusted from the position shown here to another at a 24-degree tilt. Location on the descent stage is quad 2, below the EASEP/ALSEP compartment and where the RTG cask is stored. *Author*

The MESA hinged compartment swings down once the lever is pulled, resting on the curved pipe seen by the numbered lanyard tags. Those are pulled in a sequence to access equipment for EVA activities. *Author*

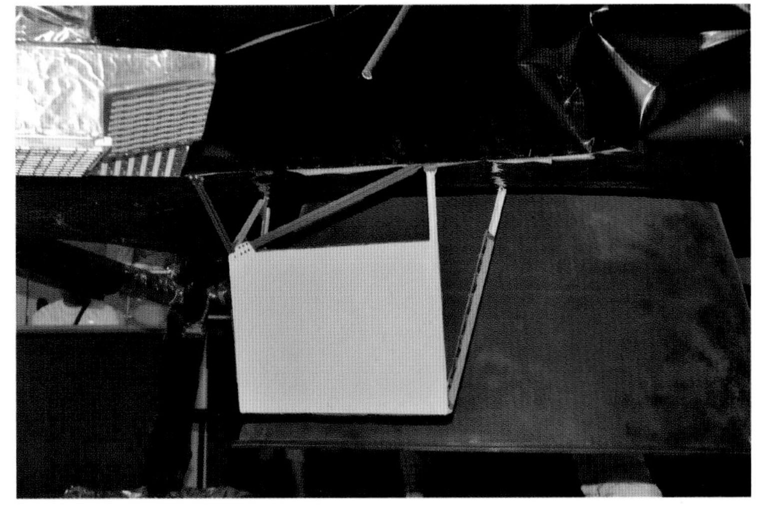

Between the radar and descent engine nozzle is this heat shield, which protected the radar from excessive heat during descent. This is on LM-2, missing on LM-9 at KSC. Both the radar and nozzle are visible at this angle. *Author*

LM-2 view of the EASEP/ALSEP bay. The astronaut unlocked the access doors, then pulled lanyards to open the bay and lower the equipment to the surface. It was then moved to be assembled. RTG truss is missing in this view. *Author*

Two cannon plugs seen here on landing leg of the +Y axis were connected to the electrical umbilical inside the SLA fixture that the LM rode inside until the CM pulled the LM free, disconnecting the plugs. *Author*

LM-9 at KSC has this truss to the left of the equipment bay, for the radioisotope thermoelectric-generator plutonium fuel cask. It was lowered by the astronaut and placed inside the ALSEP before transporting to a deployment site. *Author*

Quad bay 3 contained a supercritical helium tank. This bay was located on the opposite side of quad 1, which on J-series missions carried the lunar rover. A fairing seen here shows the tank location. *Author*

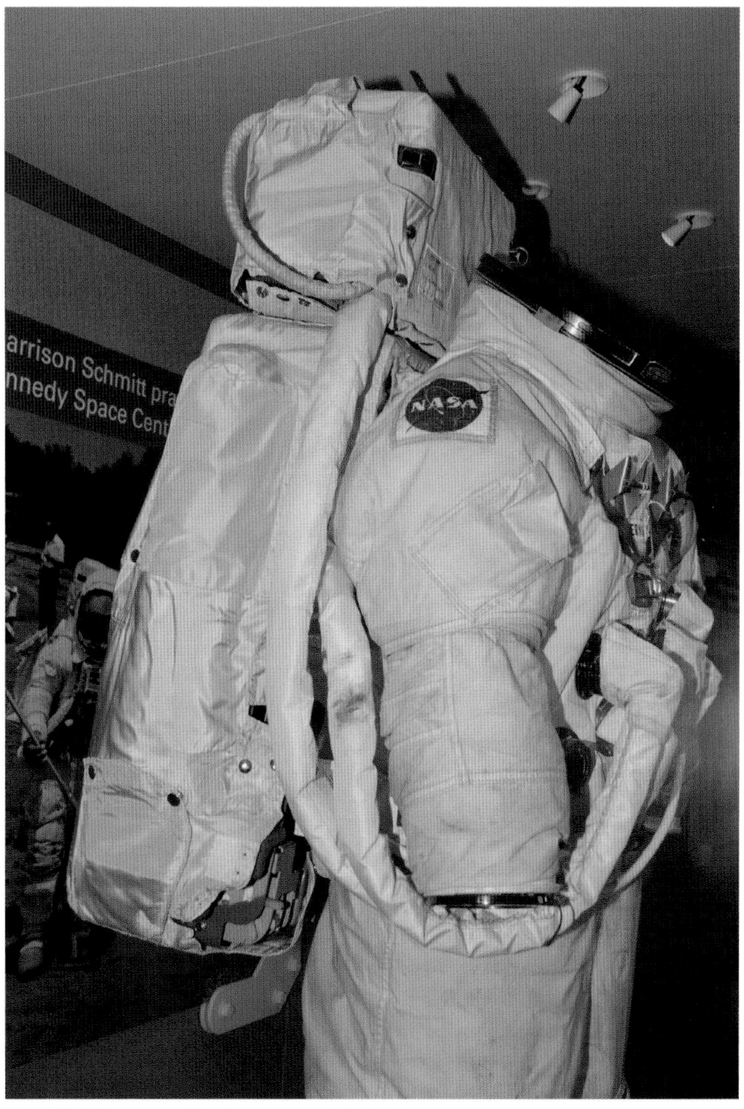

Moonwalks would not have been possible without extravehicular mobility unit (EMU) suits. The image to left is the actual moon suit worn on the Apollo 14 mission by Alan Shepard, still stained with soil from his surface activity. The red stripes identified him as mission commander. The right photo shows the training suit worn by Eugene Cernan in preparation for the Apollo 17 flight. The portable life-support system (PLSS) backpack provided the astronaut with oxygen, kept him cool, and protected him from radiation and micrometeorites. It was fitted with a radio, so while working on the lunar surface there was constant communications to receive and transmit voice messages. A short VHF antenna was raised above the oxygen purge system, the upper unit of the backpack, not seen in this photograph. *KSC, November 2011 by the author*

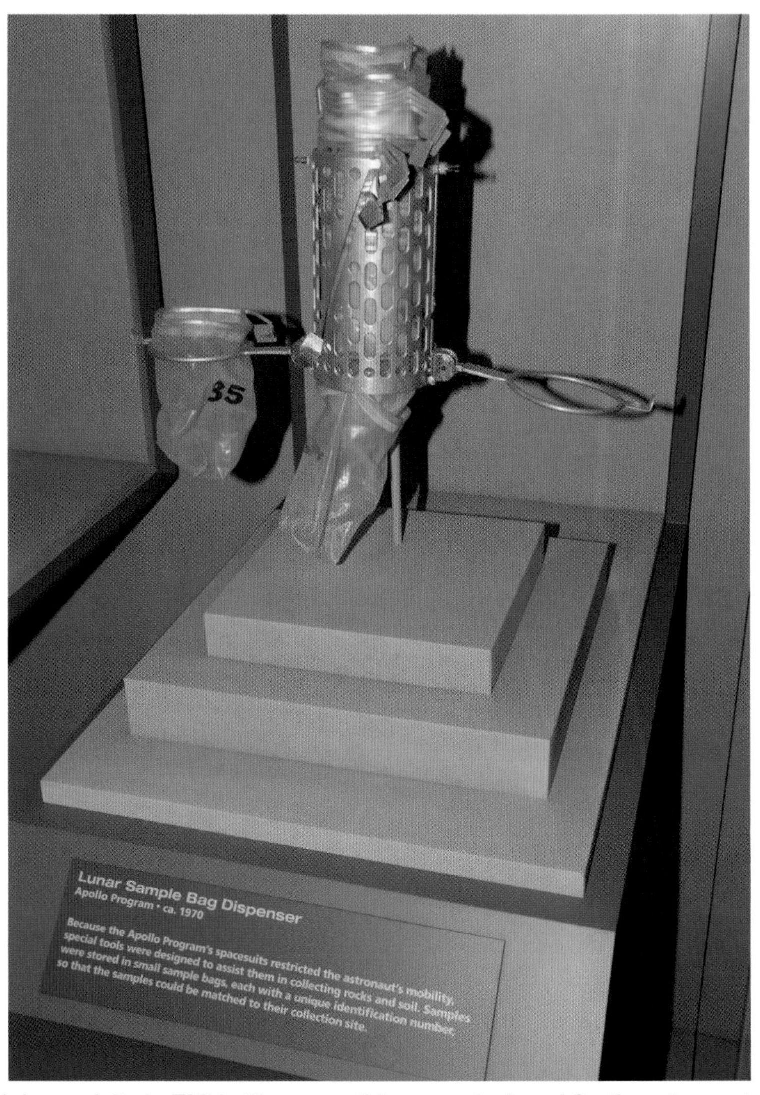

To provide the astronauts maximum operability while exploring during their lunar visits in EMU attire, everything was designed for them to meet objectives as efficiently as possible, from pulling handles or lanyards that deployed equipment, to using tools and wheeled devices such as the mobile equipment transporter or lunar roving vehicle. At the National Air & Space Museum, in the left picture (taken in July 2013), a visitor pauses to read about some of the hand tools used on the moon. Seen here are a contingency sampling tool, used to quickly obtain soil in case the EVA was terminated early. At center is a scoop/tong device, a "scong" used to grab samples without bending over. A dedicated scoop is to the right. The right-hand photo from the collection of tools at Kennedy Space Center in November 2011 shows a sample bag dispenser, again designed for the crew to use with a minimum amount of effort while an EVA was underway. *Both photos by the author*

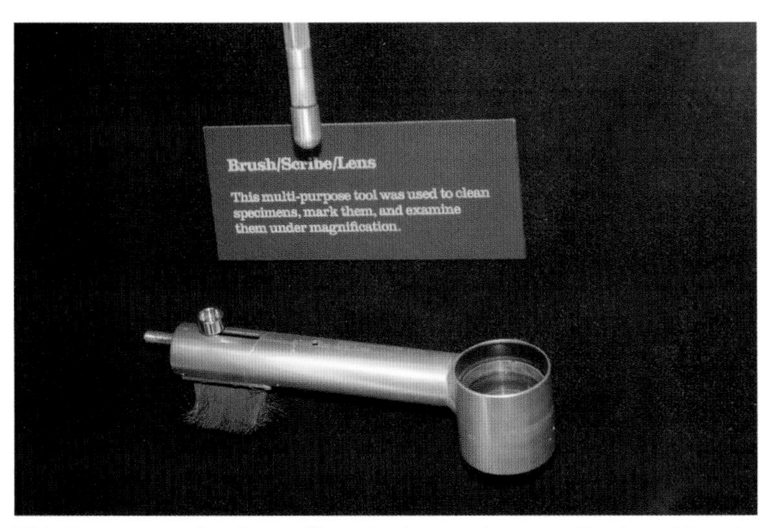

This is an example of a multiuse tool, easy to use with gloved hands. Made of aluminum for light weight and compactness. Samples were given a cursory examination by the astronauts before being bagged and cataloged. *Author*

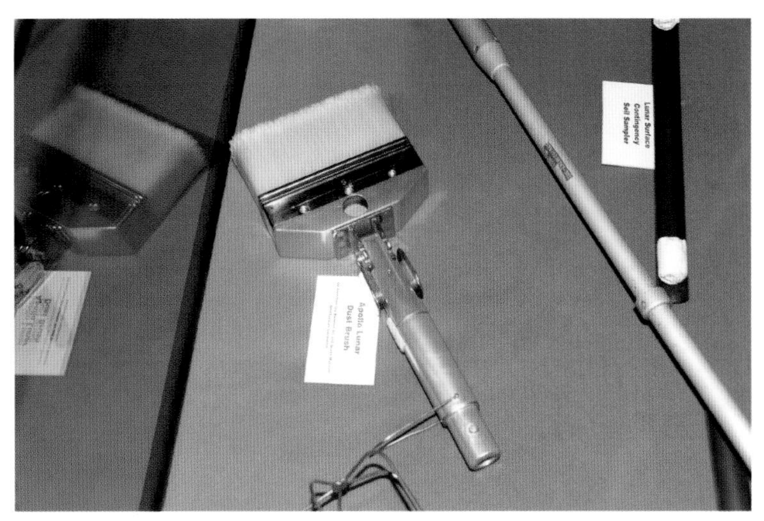

The dust brush got plenty of use during EVAs. It was essential for minimizing the accumulation of dust on everything as a moonwalk progressed. The fine powder had a gritty, abrasive texture, necessitating constant cleaning and checking of gear. *Author*

These containers stored moon rocks and soil samples. Inside was a steel-mesh liner to keep samples from being damaged during reentry and splashdown. They were not opened again until delivered to the Lunar Receiving Laboratory. *Author*

A Hasselblad camera, used for training by Apollo astronauts, used medium-format film. A chest bracket could be used to attach the camera to the EMU suit, or it could be handheld. Color as well as black-and-white film rolls were exposed. *Author*

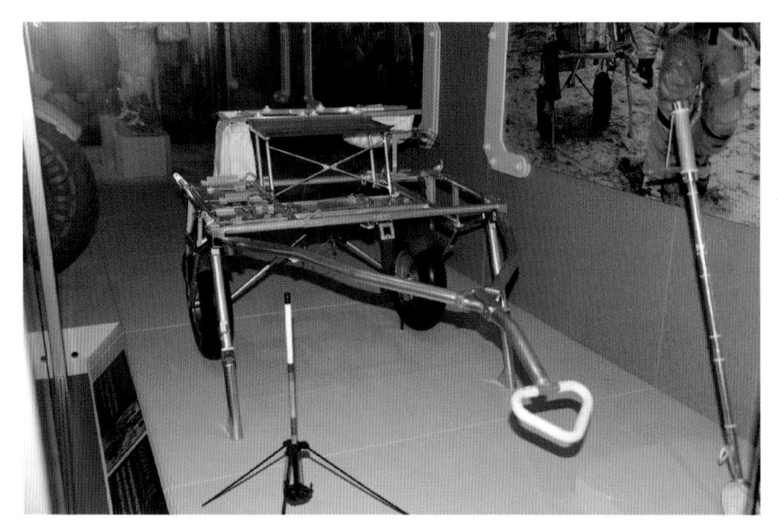

Apollo 14 used the MET—Mobile Equipment Transporter—or lunar "rickshaw," providing the ability of walking farther to explore more area. It carried all the tools, cameras, and gathered samples. *Author*

At the front of the LRV was this color television camera that could be remotely operated by mission control to observe EVA events and pan/zoom to view the lunar landscapes. The crew had to align the high-gain antenna at each stop to enable remote camera use. *Author*

The Seattle Museum of Flight has this Lunar Roving Vehicle (LRV) replica displayed. The high-gain antenna, camera, instrument panel, and tool rack all are easily identifiable. It was battery-powered, with 8 miles per hour being the top speed. *Author*

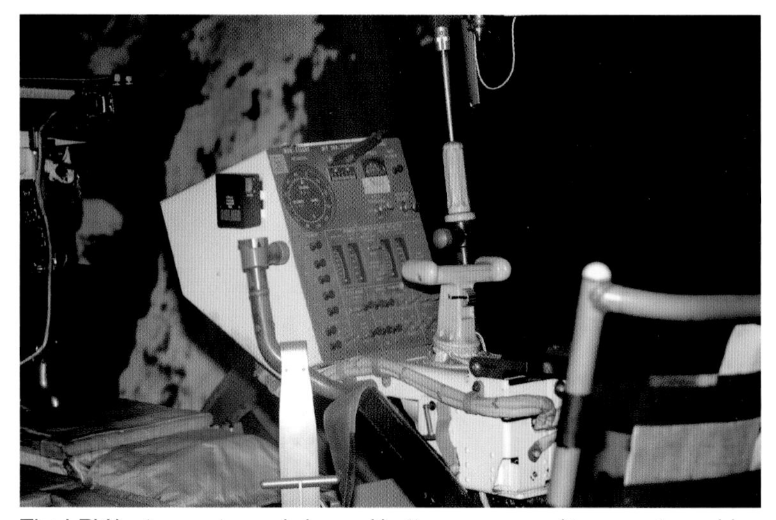

The LRV instrument panel showed battery power and temperature, drive wheel power, and speed. Round gauge showed range, distance, and bearing from the LM. Black component showed rover roll and pitch inclination. Aluminum tube was mount for low-gain antenna. *Author*

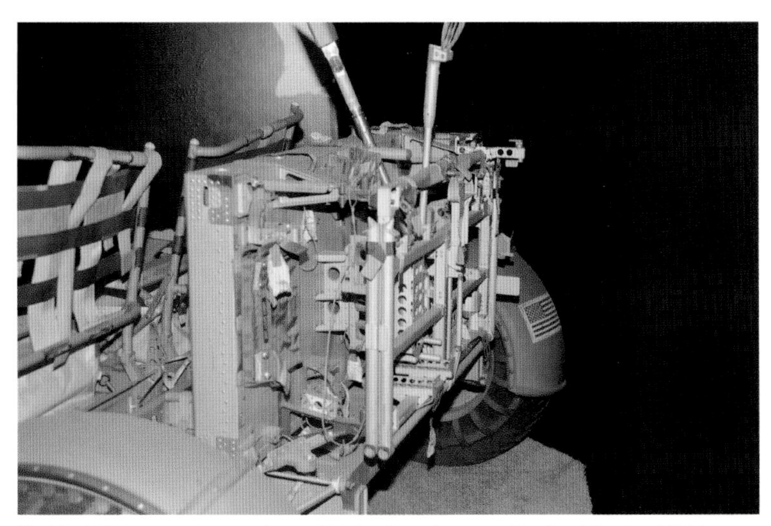

Behind the seats was where the tool rack was attached to the LRV, containing everything the astronauts would use for photographing, recording, and cataloging samples as they roamed the lunar landscape. *Author*

This is an EASEP passive seismic experiment package (PSEP) unit. It has two banks of solar panels, and two small RTG power sources providing heat. They are within the small cylinders close to the white antenna mast. *Author*

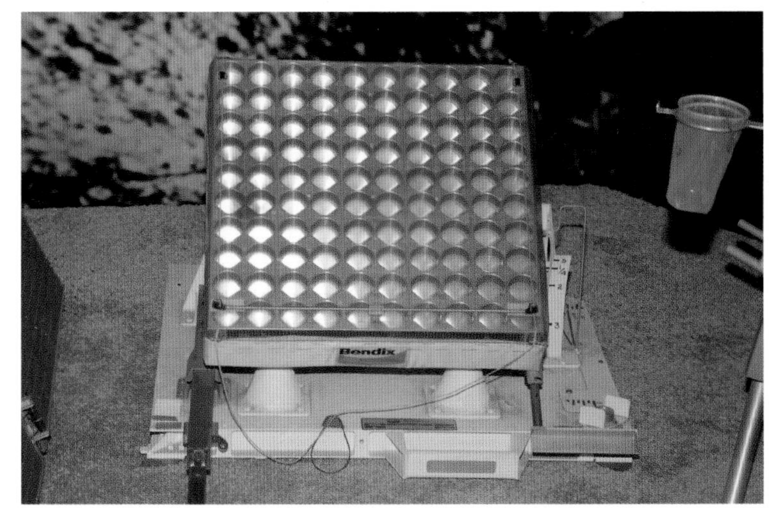

The LRRR, or laser-ranging retroreflector, was left at Apollo 11, 14, and 15 landing sites. It enables earth-to-moon distances to be measured within 15 centimeters. One hundred prisms of fused silica had no moving parts. *Author*

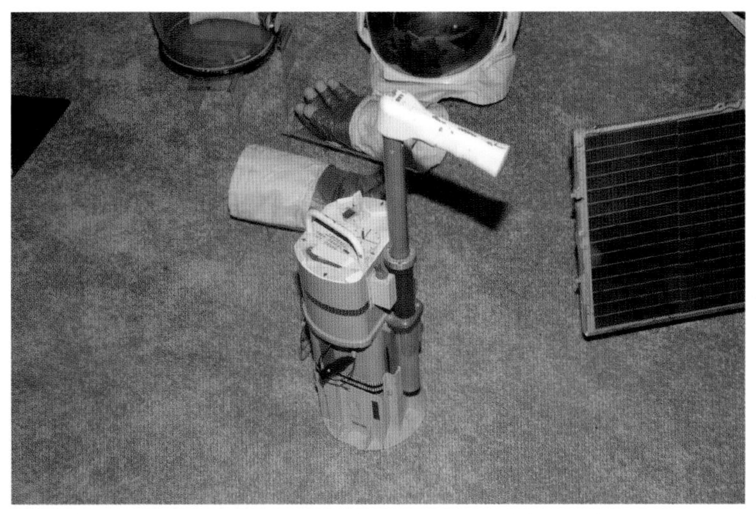

This stereo camera was used to obtain images of the surface that could be viewed with a stereoscope. It was placed on the surface as shown, handle trigger pushed. The resulting photo enabled a 3-D effect when examined later. *Author*

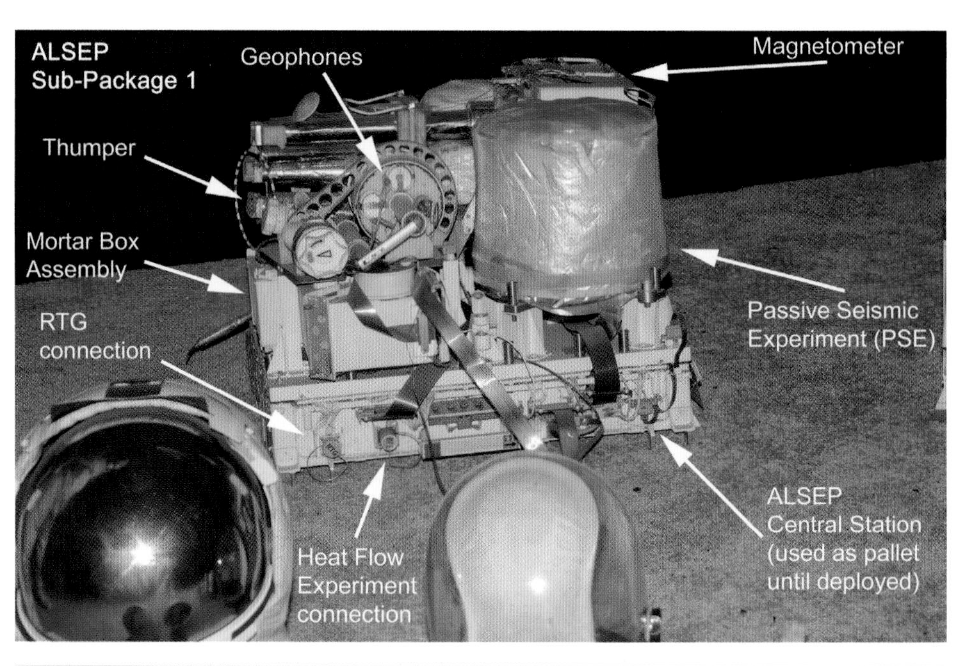

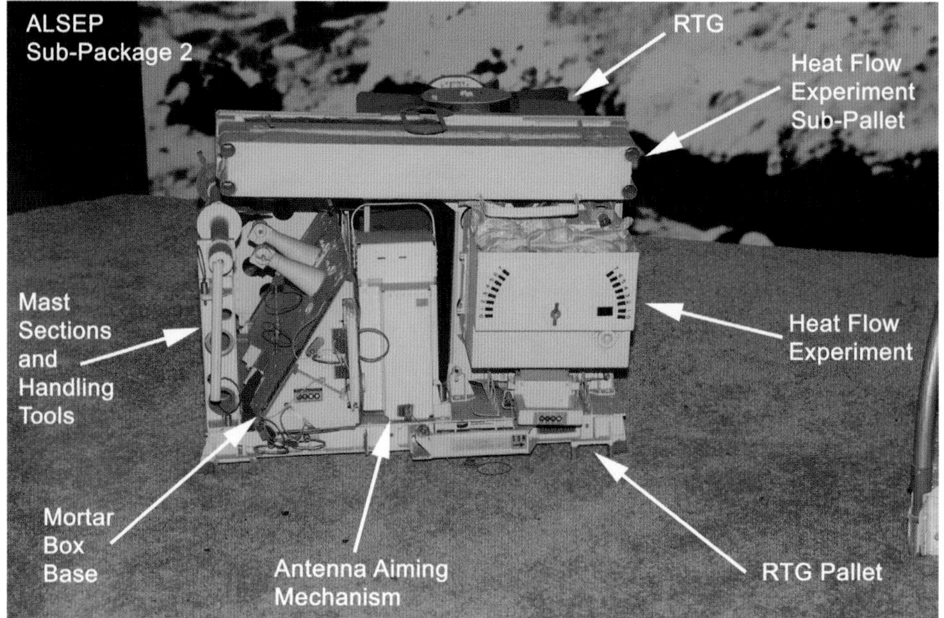

ALSEP instruments were installed on pallets that were unloaded off the LM descent-stage quad 2 bay once the first lunar EVA began. After both pallets were on the surface, the RTG cask was lowered. The plutonium element was removed, then inserted into the RTG generator. The astronauts could then carry both pallets by attaching between them an antenna mast section that also served as a "barbell" handle, or with the LRV it all could be driven to a suitable location for deployment. Sub-Package 1 contents included the geophones used to detect seismic waves, the "Thumper" used by astronauts to create seismic waves, and the passive seismometer for long-duration monitoring of lunar geologic events as well as planned impacts by items of known weight and size, such as the spent S-IVB third stage and the LM ascent module. The pallet base was used after everything was removed from it as the central station, where all the power distribution and electronics that ran the deployed equipment were installed. A mortar box had four rocket-propelled charges that were fired off once the astronauts were gone. It created waves at shallow depths that provided data on the subbearing strength and physical composition of terrain down to 500 feet. The Heat Flow Experiment (HFE) transmitted data sensed by two heat probes that were inserted into 20-inch-deep holes drilled into the lunar surface. The magnetometer was used to read measurements of the moon's magnetic field at the surface. Two plugs marked RTG and HFE were where the astronauts would insert electrical connectors for the RTG to supply power, and for readings of the HFE, to be relayed to Earth for analysis via the central station. Sub-Package 2 contained the mortar box base that attached to the box itself, the mechanism used to align the ALSEP S-band data relay antenna, antenna mast sections, and various handling tools for the experiments, and the HFE components. The RTG SNAP-27 plutonium-238-type generator was the black cylinder partially seen on the background of the loaded pallet, with eight heat-rejection fins that powered the whole ALSEP array. It remained fixed to the pallet, which was the base for the unit. ALSEP contents / pallet load-outs varied depending on goals assigned to each mission. *Both photos, National Air & Space Museum, taken by the author in July 2013*

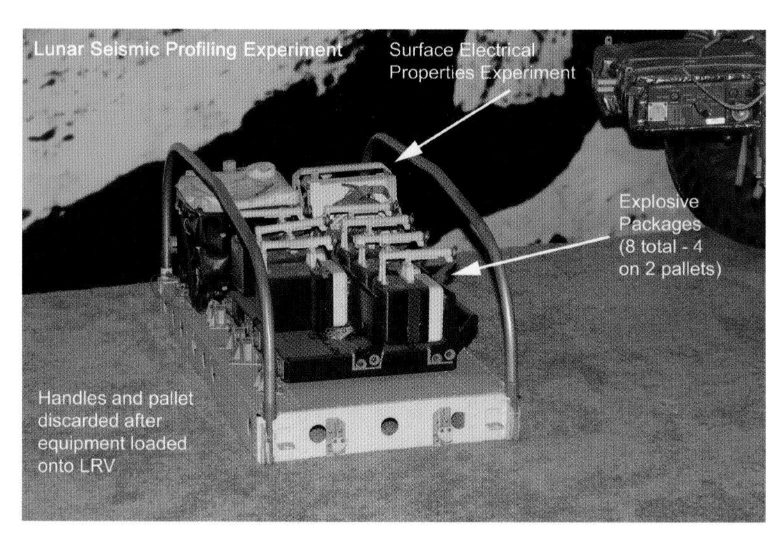

Apollo 17 carried this pallet loaded with eight explosive charges and the Surface Electrical Properties Experiment (SEPE). All of this was secured on the LRV. Each was placed on the basis of distance and strength of waves it could generate.

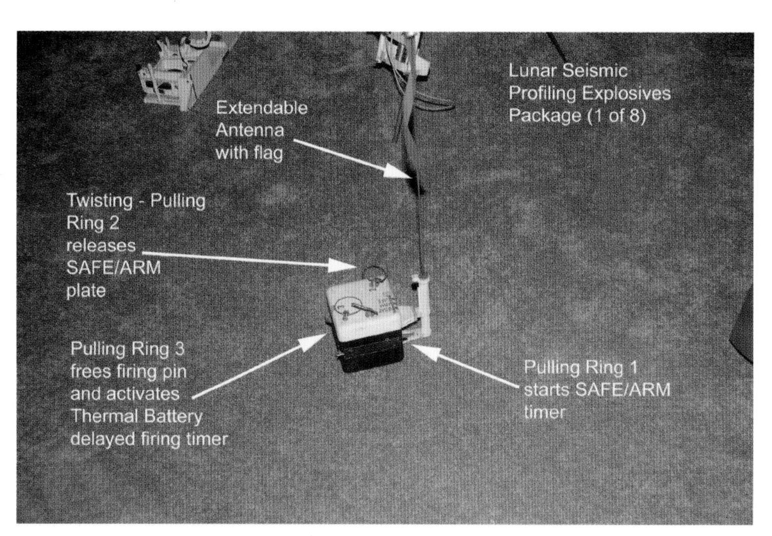

A lunar seismic profiling (LSP) explosive charge shown set in position. Arming it included pulling the three rings to begin the sequences. They were fired remotely from Earth, long after the astronaut EVAs had concluded.

Powering ALSEP arrays was a SNAP-27 plutonium atomic element. Unit was installed on the side of an LM. Contents were taken from cask and inserted into device at left. Heat decay was converted into electricity to power all instruments.

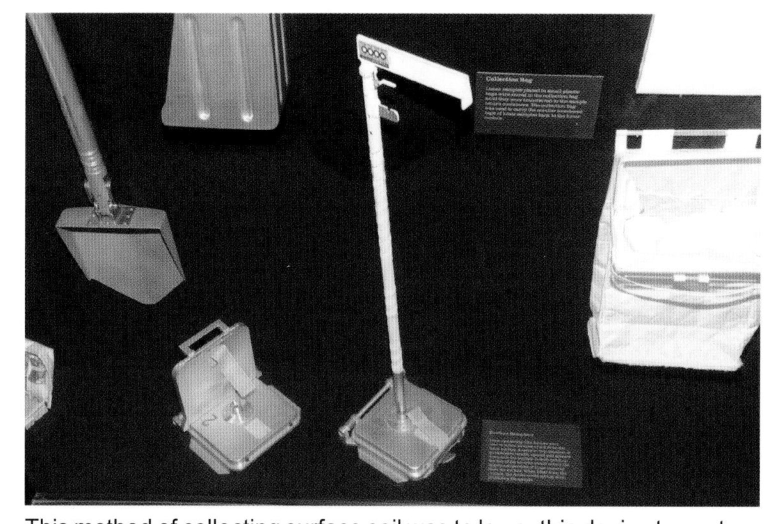

This method of collecting surface soil was to lower this device to capture material. Soil clung to a cloth padding. Lifting the sampler released a spring latch, closing the unit. Another one could be used on the end of the pole if desired. *This page, National Air & Space Museum, taken by the author in July 2013*

Yet another Apollo-specific innovation was the docking system between the CM/SM and LM. The top photo shows almost all the related components. At far left in the top picture is the tan hatch and cone-shaped drogue that was in place inside the lunar module. The maroon-shaded docking ring was a permanent part of the CM. Missing in this display is the LM docking ring, which was a permanent part of the ascent-stage top section. The hatch to the right of the CM docking ring was an ablative design that sealed the CM capsule. The docking probe had at the front of the movable arms a cone that contained three evenly spaced spring-loaded latches. As the CM slowly approached the LM, the probe was fully extended, as depicted in the bottom image. Both pilots used visual docking targets to aid in proper alignment, firing thrusters if necessary to make corrections as the distance between the two closed. The probe, once centered in the drogue, achieved a "soft dock" once the three latches snapped open inside the drogue hole. The probe then retracted, pulling the two spacecraft together. Once the CM and LM docking rings came into contact, twelve CM docking-ring latches engaged. This phase was "hard dock," and with this achieved, air pressures were equalized, with an electrical connection between the two established. The docking mechanisms could be removed and stowed to allow for the connecting tunnel to be unobstructed for movement of crewmen between the two modules. After the rendezvous in lunar orbit following EVA activities, the crewmen of the LM could pass into the CM the haul of samples and equipment as well as move themselves into the CM, in preparation for undocking the LM ascent stage. That would be sent crashing into the moon as a way of checking seismic experiments. With docking maneuvers at that point concluded, the probe and drogue units were placed inside the LM ascent stage, no longer needed. If docking by this method malfunctioned, the backup option was to conduct an EVA between the two modules to reunite the astronauts. This exhibit is on public view at the Huntsville Space & Rocket Center. *Photos taken by the author in October 2016*

The Hazy Center has the Mobile Quarantine Facility (MQF) used on Apollo 11. The astronauts were kept in isolation after their return as a precaution against transmitting any moon-borne germs. They stayed inside for nearly ninety hours. *Author*

The MQF had filtration and communications installed. Total complement was five: the three astronauts plus a physician and technician. This view of the MQF pallet shows some of the exterior supporting gear. *Author*

The Space & Rocket Center has the Apollo 12 MQF. It was on the recovery aircraft carrier and could be lifted by crane. It was also airlifted to Texas, where a more spacious isolation facility was available. *Author*

This is a view of the Apollo 12 MQF common area, where meals could be eaten, board games could be played, and reading could be done. There was a kitchen, bunks for sleeping, and a lavatory. Four were built, and after Apollo 14 this procedure was deemed not necessary. *Author*

Chapter 5

SKYLAB AND ASTP

The last Saturn V rocket placed the Skylab Orbital Work Station into orbit. It was nearly disastrous, since the meteoroid shield and an entire solar panel were torn off during launch. The Skylab 2 crew were able to repair the station enough for them to stay, and for two follow-on crews to inhabit it and conduct experiments to gather data on long-duration space effects on the human body. Sadly, the Skylab was lost before a space shuttle was able to boost it into a higher orbit. Two Skylab artifacts are on display, one in Washington, DC, and the other in Alabama.

The Apollo-Soyuz Test Project (ASTP) was a joint American-Soviet agreement to develop the means to dock in space and visit each other's spacecraft while conducting experiments in the interests of détente and mutual cooperation. A docking module (DM) was designed to allow a docking to occur. A full-scale ASTP exhibit with DM is at the National Air & Space Museum.

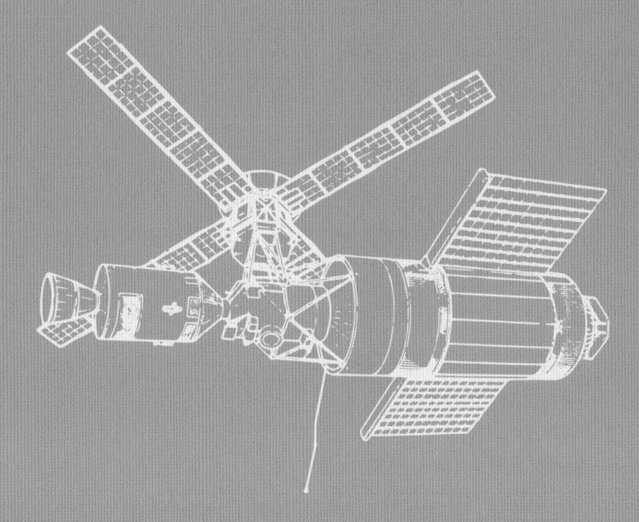

SKYLAB AND ASTP

Date	Mission	Crew	Duration
May 14, 1973	Skylab 1 / Saturn V	Skylab OWS launch	orbital workstation launched
May 25, 1973	Skylab 2 / Saturn IB	Conrad/Kerwin/Weitz	28 days
July 28, 1973	Skylab 3 / Saturn IB	Bean/Garriott/Lousma	59 days
November 16, 1973	Skylab 4 / Saturn IB	Carr/Gibson/Pogue	84 days
July 15, 1975	Apollo-Soyuz	Stafford/Slayton/Brand	test project / Saturn IB docked with Soyuz 19

Skylab 1 launch. Moments after this photo was taken, the micrometeoroid shield was torn off the workstation. One entire solar panel array followed shortly afterward. Once in orbit, the other solar panel was inoperable. The Skylab 2 launch was delayed while solutions were devised to make repairs to the station. The first visiting crew was able to erect a sunshade, which lowered high temperatures inside and freed the jammed solar array. That provided the power needed to recharge batteries and conduct experiments. *NASA photo*

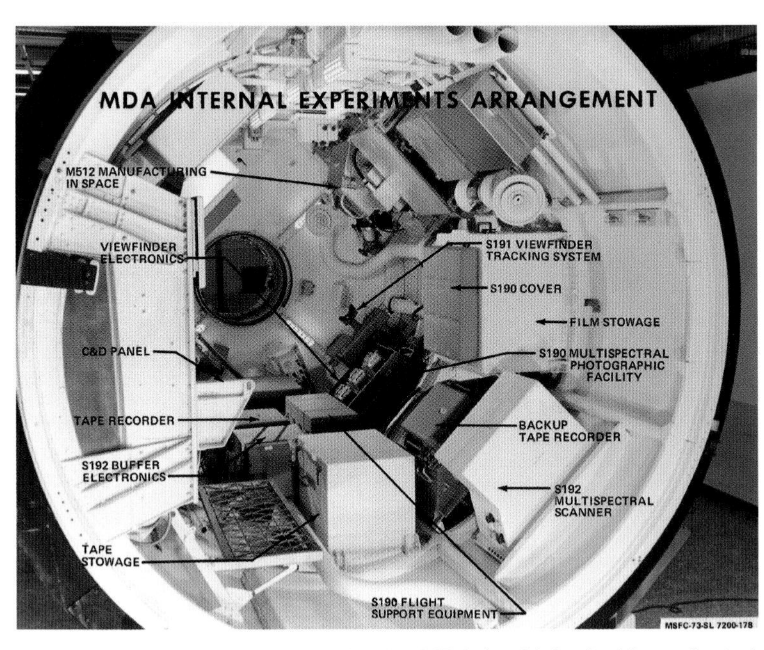

Two NASA highlighted images, showing, *at left*, the OWS (orbital work station) crew quarters, and, *at right*, the MDA (multiple docking adapter) interior arrangement. The OWS was a converted S-IVB Saturn third stage and was the largest component of the entire design. NASA drawing below shows Skylab in orbital configuration.

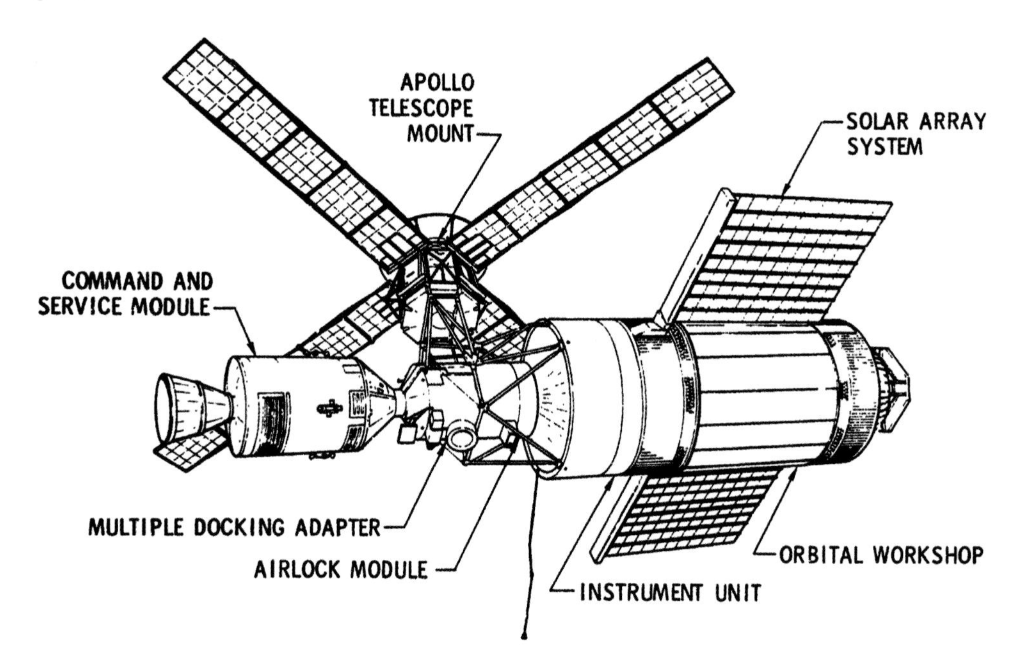

The backup Skylab OWS on display at the National Air & Space Museum. This is one exhibit modified for public access so that visitors can see the interior arrangement. One solar panel is deployed. *Author*

At the Huntsville Space & Rocket Center, a restored Skylab full-scale engineering OWS is open for viewing inside the Davidson Center for Space Exploration. This allows a detailed look at the Skylab living quarters. *Author*

Huntsville also has this outside mockup of a Skylab with an S-IVB third stage, multiple docking adapter, and CM/SM (*to right*). This is how Skylab would have appeared in orbit with a crew aboard. *Author*

This section is the MDA, with OWS to the left and CM to the right. The round access, facing camera, is the radial docking-port hatch. Missing is the suite of EREP (Earth Resources Experiments Package) items, which were externally mounted. Four oxygen tanks are at left. *Author*

Two views of the Apollo Telescope Mount (ATM) at the Huntsville Space & Rocket Center, in October 2016. It is in the deployed position; for launch it was centered over the docking hatch, seen at far left in the left-hand image. Sun observations and measurements were the prime ATM objective, with a suite of sensors dedicated to the work. Area where the mannequin is located is where an assortment of covers protect the sensors. Right photo shows arm for antenna wires, large sunshade, electrical junctions, blue EVA handrails/footrails and areas where astronauts would go on EVAs to retrieve film. ATM operations were controlled from within the MDA. The chart below shows a NASA photograph, with items numbered for identification by the author. *Two photos above by the author*

1. X-Ray Spectrographic Telescope
2. Fine Sun sensor assembly
3. XUV Spectroheliograph
4. H-Alpha No. 2 Telescope
5. XUV Spectrograph
6. UV Scanning Polychromator Spectroheliometer
7. H-Alpha No.1 Telescope
8. Rate Gyro
9. White Light Coronagraph
10. Dual X-Ray Telescope
11. X-Ray Event and Analyzer Assembly

Two views inside the National Air & Space Museum's Skylab OWS show, *at left*, an ergometer, and, *at center*, with white, partially raised shower curtain behind it and in the foreground, a rotating litter chair used to test astronauts' reactions to motion inputs in weightless conditions. During the Skylab long-duration stays in space, astronauts kept up a vigorous exercise routine, while being monitored and documented to assess humans' ability to exist in the space environment. At right is the waste management system compartment. The ability to design, operate, and integrate this vital system into the overall station was another goal of the program. *Both photos by author*

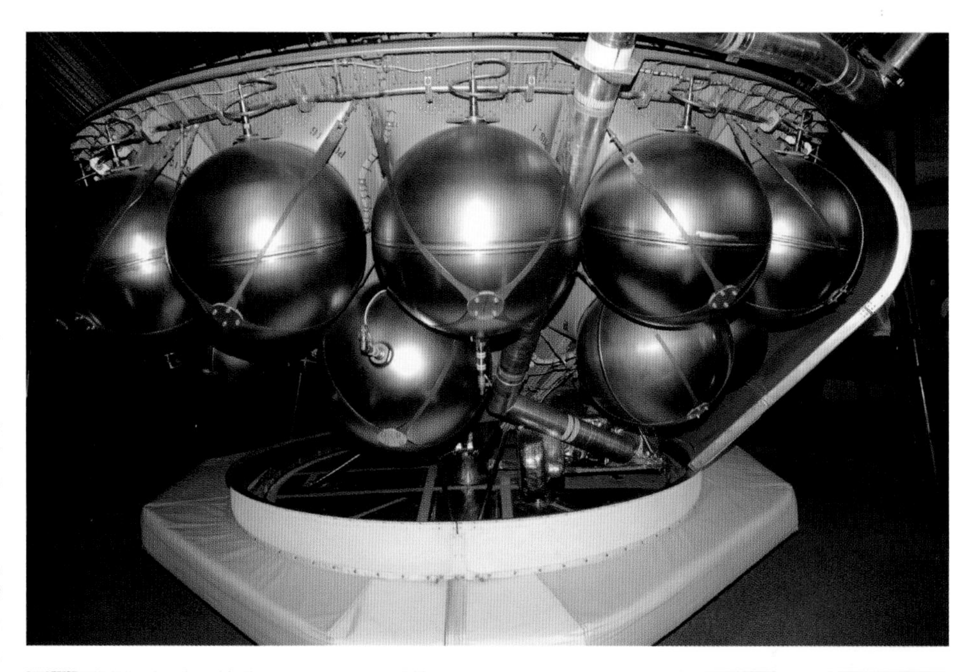

The National Air & Space Museum's Skylab exhibit, in Washington, DC, has a section of the OWS bottom fairing removed to show some of the twenty-two spherical tanks used to store nitrogen gas, used as propellant for the TACS (thruster attitude control system). These would normally be fully enclosed inside a meteoroid shield. Part of this shield is visible to the right in the top image. Silver ductwork moves heated coolant through the radiator. Components in gold foil are parts of the radiator control system. The white device at the base is the station radiator, used to dissipate heat from electronic equipment into space. Since the OWS was a modified S-IVB third stage, the light-green conical thrust structure for a single J-2 rocket engine is partially visible behind all the items unique to Skylab. Skylab was intended to be kept in use after the three scheduled astronaut visits had concluded. One early objective of the follow-on space shuttle program was to launch the orbiter with a teleoperator retrieval system (TRS), which would have been attached to the Skylab axial docking-port section. This was capable of raising the station into a higher orbit, but the orbit decayed over time to the point that it reentered the atmosphere before the first shuttle flight occurred. One of the six high-pressure oxygen storage tanks aboard the OWS, among other items, survived the atmospheric friction and landed on the ground in Australia. This was brought back to the US and is displayed at the Huntsville Space & Rocket Center. The steel tank was wrapped in layers of composite fiber sheets to provide more strength, as opposed to making a tank of thicker steel, which would have added weight. The layers making up the composite exterior are visible in the bottom photograph. *Both photos by author*

The Apollo-Soyuz Test Project (ASTP), flown in July 1975, was a cooperative effort between the United States and the Soviet Union to rendezvous and dock in earth orbit. A docking module (DM) was jointly designed to mate the two dissimilar spacecraft together by way of a three-leaf androgynous docking system. It also allowed for crews to adjust to different cabin pressures as they visited each other's capsules. The Apollo crew consisted of Thomas Stafford, Deke Slayton, and Vance Brand. The cosmonauts were a crew of two, Alexei Leonov (the first man to walk in space) and Valery Kubasov. In addition to the symbolic significance of the mission, the ASTP also conducted many experiments while docked and undocked. The DM contained a multipurpose furnace and ultraviolet absorption components. The mission cost $245 million; in 2010 dollars the amount would have been $1 billion. DM in photo was a flight backup. *Display at the National Air & Space Museum, photo by the author*

ASTP DM side view, showing oxygen and nitrogen tanks. These were needed to pressurize the DM several times during crew movements from Apollo to Soyuz. Pressure difference was 5 psi in Apollo, 14.7 psi in Soyuz. *Author*

DM weighed 4435.7 pounds; its length was 10 feet, 4 inches, with a diameter of 4 feet, 8 inches. Black rectangle contained the MA-059 ultraviolet absorber/spectrometer. Oxygen and nitrogen tanks had insulation and cover when used. *Author*

Apollo CM/SM end of ASTP exhibit. Service module white panels are radiators. Maroon-shaded area with dark circle is fuel/defuel line valves for SM engine. Apollo pulled DM out of S-IVB stage, as it had done when removing LM. *Author*

Soyuz replica at the NASM represents Soyuz-19. Section attached to DM is the orbital module; next section is the descent module. Section with solar panels is the service module, functioning the same as the SM on Apollo. *Author*

Chapter 6

SPACE SHUTTLE AND INTERNATIONAL SPACE STATION

The Space Transportation System / Space Shuttle created reusable components in the form of solid rocket boosters, and an orbiter that took off like a rocket and landed like an aircraft onto a runway. It could carry items into orbit and deploy them, or retrieve satellites and bring them back for repairs and eventual release to continue their missions. It was a major asset in the assembly and outfitting of the International Space Station, or ISS. For all of its advances and groundbreaking technology, the space shuttle fleet proved hard to maintain and was fragile and expensive to operate. Three surviving orbiters are on exhibit in California, Florida, and Virginia. From 1972 until 2011, the space shuttle program cost $196 billion.

The International Space Station is an international partnership of sixteen nations, spending a combined total of $120 billion. Pressurized volume inside the ISS is 32,333 cubic feet (about the same interior space as a 747 airliner). It is the third-brightest object in space, after the moon and Venus. About 8 miles of wire connect everything into the station's electrical systems. It completes sixteen orbits of Earth each day, weighs 450 tons, and is the largest orbiting man-made object ever assembled. It has been in low earth orbit for twenty years and four months as of March 2021.

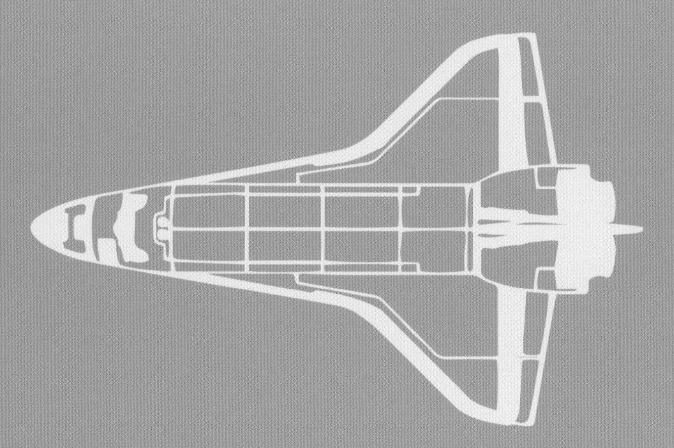

The Space Transportation System (STS) concept had its roots in substantive studies going back as far as the 1940s. By the 1970s, many practical designs were offered by the aerospace industry as to what such a vehicle could look like: an orbiter with no tail fin, upswept wings of straight and delta shapes with a launching platform, internal/external rocket boosters, or enough takeoff power to go from a runway into orbit. Several models of differing sizes and propulsion arrangements are shown in the two photographs to the left, taken at the American Space Museum in Titusville, Florida, in January 2019 by the author. Three space shuttle orbiters are currently on display in the United States. *Atlantis* is in a dedicated gallery at the Kennedy Space Center visitor complex. *Discovery* is the centerpiece of the James S. McDonnell space hangar at the Hazy Center. *Endeavour* is in California, at the Samuel Oschin Air and Space Center. A total of five orbiters were built; two of those, *Challenger* and *Columbia*, were lost in accidents. The shuttle program achieved several goals, from the launching/recovery and servicing of satellites such as the Solar Max and Hubble Space Telescope, to delivery into orbit and assembly of major components of the International Space Station (ISS), to advancing the studies of human abilities to live and work in microgravity. Improved EVA techniques and trying out new equipment such as the remote manipulator system or robotic arm were put to use, and techniques were perfected: new engines, ceramic tiles able to absorb atmospheric reentry friction, a payload bay that could be custom-arranged for payloads weighing up to 65,000 pounds, and reentering the atmosphere and landing on a runway like an airliner. All were remarkable engineering successes put to practical use. From lake bed landings, orbiters were towed into mate/demate structures, lowered onto 747 SCA (shuttle carrier aircraft), and flown back to KSC. Landings that ended at the KSC SLF (shuttle landing facility) involved a post-landing towing into an OPF (orbiter processing facility) and made ready to fly again and again, along with pairs of reusable solid-rocket boosters. All these milestones and much more were achieved from the launch of *Columbia* in April 1981 until July 2011, when the last flight, conducted by the *Atlantis* orbiter vehicle, occurred.

Atlantis orbiter vehicle (OV-104) top photo, as displayed at the Kennedy Space Center in Florida. Payload bay doors open, arm extended. Orbiter is tilted with the port (left) wing down at an angle of 43.21 degrees, the numbers symbolic of a countdown, where visitors can easily see the payload bay section and equipment inside. A wide spiral walkway provides wonderful views of the vehicle from multiple angles. Various other items of shuttle-related equipment are also exhibited in the facility. *Atlantis* flew five times on missions to launch satellites for the Department of Defense (DoD) and carried the Magellan, Galileo, and Compton Gamma Ray Observatory into orbit. It also visited the Russian Mir space station seven times, as well as flying the last shuttle mission. *Discovery* (OV-103) was originally intended to be based at the Vandenberg AFB Space Launch Complex 6 in California, conducting high-inclination launches of mostly scientific payloads. After the *Challenger* accident, it was flown to KSC and used on NASA missions, including the launch of the Hubble Space Telescope (HST), and also was used on two HST servicing flights. It also flew both return-to-flight events following the loss of *Challenger* and *Columbia*. She was the first to dock with the ISS. Of all the orbiters, it flew the most missions, with thirty-nine flights accomplished. *Discovery* is displayed with landing gear extended and payload bay doors closed, with a remote manipulator system arm alongside it for viewing. The orbiters themselves were 122.17 feet in length, with a wingspan of 78.06 feet, and stood 56.58 feet high. Empty weight averaged 150,000 pounds. Payload bay length was 60 feet, capable of accommodating items of up to 15 feet in diameter. Nose sections of all orbiters as well as the wing leading edges were of a reinforced carbon–carbon material, not TPS (thermal protection system) tiles as seen elsewhere. This was due to the amount of heat those areas had to absorb during reentry. Each orbiter had three SSMEs (space shuttle main engines) in the aft section that were used during launch. These were augmented by the pair of solid rocket boosters (SRBs). An extensive array of maneuvering thrusters were installed for positioning the vehicle in orbit. Three fuel cells provided electricity and water. Three APUs (auxiliary power units) provided hydraulic pressure. *Photos by the author*

Atlantis forward RCS (reaction control system) module section housed fourteen primary and two Vernier thrusters. Also within the section are the fuel supply, lines, and valves to operate everything. *Author*

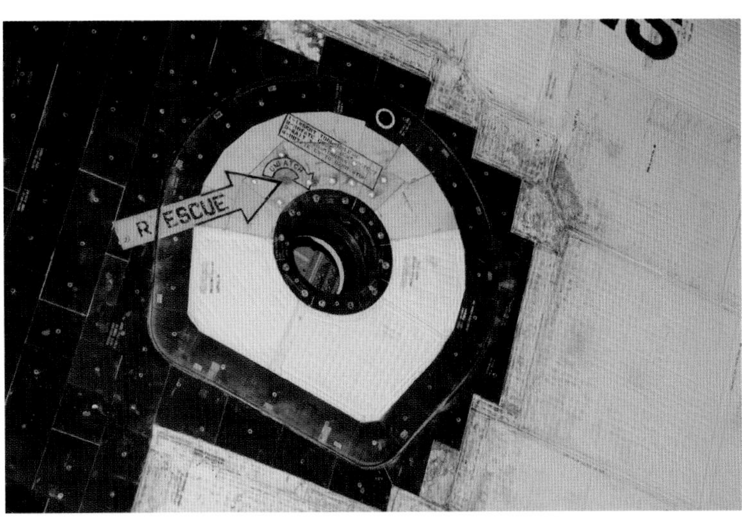

The normal crew entry/egress hatch on the left side was 40 inches in diameter and weighed 294 pounds. Rescue instructions of how to open the hatch from outside the orbiter are highlighted in yellow on *Atlantis*. *Author*

Nosewheel tires are hinged to fall aft when extended, assisted by the airflow just prior to touchdown 300 feet above the ground. Steering is by an electrohydraulic actuator. Tire pressure is 350 psi of nitrogen. *Author*

Crew compartment windows on *Atlantis*, aft of the flight deck. These were used to provide visibility for docking maneuvers. Red triangle indicates the left window section can be jettisoned. This gives the crew a secondary exit from the orbiter. *Author*

Two *Atlantis* photos, showing, *at left*, the pair of star trackers just forward of the flight deck windows on the left side, used to orient the Shuttle during orbital phases of flight. They also feed data to the three IMUs (inertial measurement units). Upper unit is the Z-axis sensor; lower one images in the Y-axis. Doors are closed as they would be on launch and reentry, opened while in orbit. The right-hand image shows two payload bay cameras, which could be remotely operated from the ground. Large item in bay is the ODS (Orbiter Docking System), to allow docking with the Mir space station. This consisted of an orbiter airlock, an external truss to support the upper device, and the APDS (androgynous peripheral docking system), the gray/silver assembly. The unit cost $95,200,000 to design, manufacture, and install. It was 15 feet wide, 6.5 feet long, and 13.5 feet high and weighed 4,016 pounds. This one, mounted in the payload bay, is a replica. *Author*

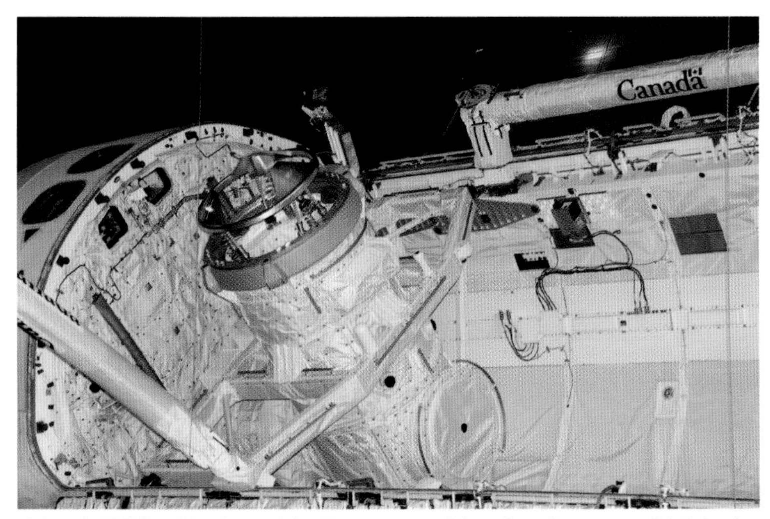

Atlantis ODS with view of the large, round hatch at bottom, which is the orbiter airlock. What looks like the arm at top is an extension called the orbiter boom sensor system, for viewing thermal protection tiles on the belly area. *Author*

Atlantis open left payload bay door, with radiators used to rid the orbiter of heat. This door always opens last and is the one closed first. The doors are 10 feet wide. The left door weighs 2,375 pounds; radiators add 833 pounds. *Author*

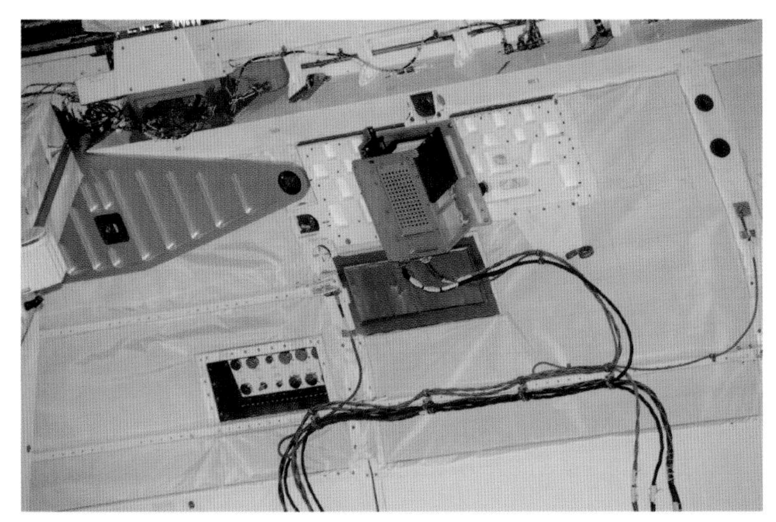

Along the right *Atlantis* payload bay wall is this container that the Picosat (see page 62) was sent into orbit from. Small satellites are easy to build and launch and are tailored to specific requirements. *Author*

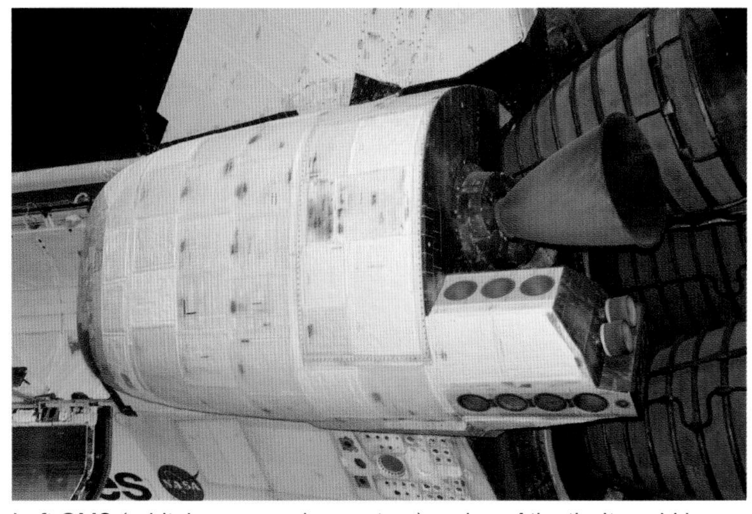

Left OMS (orbital maneuvering system) pod on *Atlantis*. It could be removed as a unit for servicing or to be swapped out. Engine was used for orbit insertion, to circularize an orbital path, and to deorbit. *Author*

Discovery right wing, a unique "double-delta"-shaped leading edge. Also seen are the inboard/outboard trailing-edge elevons, used in the atmosphere. Movement is up 40 degrees, down 25 degrees. *Author*

Discovery undersides show open pair of external tank (ET) umbilical access doors. These are closed once the ET separates from the orbiter during launch. They are tiled and are 50 square inches in area each. *Author*

Discovery main landing-gear assembly with bay door open. Exterior of door is also tiled. Wheel brakes are of beryllium rotors / carbon stators. They can absorb 36.5 million foot-pounds of energy for normal stopping. *Author*

Left-hand ET umbilical connections on *Discovery*. This is where SSME fuel supply of liquid hydrogen / liquid oxygen is fed into the orbiter. The ET, once discarded, is the only major component not reused. *Author*

At the entrance to the *Atlantis* exhibit at Kennedy Space Center are a pair of solid rocket boosters (SRB) attached to an ET. In the left photo are the attach points for the orbiter, the bipod (*at the top*), and the two ET fuel and electrical connections (*at bottom*). ET top section is the liquid oxygen tank. An intertank structure (seen in the area of the cone-shaped tops of the SRBs) fills the void between the bottom of the oxygen tank and top of the liquid hydrogen tank. The SRBs are solid-fueled and provide most of the liftoff power to achieve orbit. Once lit, they cannot be shut off. Each booster is assembled in segments, with the solid-fuel amount designed to provide full power at liftoff. Thrust is reduced to around 70 percent fifty-five seconds after launch, to prevent overstressing the vehicle during the phase of maximum dynamic pressure, known as the Max-Q event. The top SRB sections and bottom skirts contain separation motors that push each booster away from the ascending orbiter/ET once they burn out. Both SRBs fall to the ocean by parachute, to be recovered and reused. *February 2014 by the author*

Atlantis belly view, with blue mounting bracket shown. Photo shows the closed position of the pair of ET doors. When closed, they are pulled into place to seal the area tightly, preventing reentry heating damage. *Author*

Aft *Atlantis* OMS pods have arrays of thrusters. Seen here are nine of the type R-40A primary thruster units. Two TPS blankets between thruster rows are of AFRSI (advanced flexible reusable surface insulation). *Author*

Outboard-facing *Atlantis* aft OMS pod thrusters are of four R-40A units. Each rated at 870 pounds thrust. The small one (*at right*) is one of the Vernier thrusters, type R-1E-3. It is rated at 25 pounds of thrust. *Author*

This *Atlantis* aft fuselage left-side view shows the access door next to the NASA logo to enter into the aft section. At upper right is the interface used by crews after landing to defuel the orbiter of hazardous propellants. *Author*

Atlantis aft section, showing where the space shuttle main engine (SSME) and OMS and RCS components are located. The base of the tail fin had a housing installed for a drag chute, to shorten landing rollouts and to reduce brake and tire wear. *Author*

An RS-25 or SSME at the *Atlantis* exhibit. Each second, it consumed 889 pounds of liquid oxygen and 146 pounds of liquid hydrogen at a 100% power setting. Each engine in a vacuum produced 470,000 pounds of thrust. *Author*

A prototype partial OMS engine at the Seattle Museum of Flight. The engine without nozzle weighed only 305 pounds but could provide 6,000 pounds of thrust. The fuel it burned was nitrogen tetroxide oxidizer and monomethyl hydrazine fuel. *Author*

This SSME, shown at the Huntsville Space & Rocket Center, at the angle photographed shows the black MEC (main engine controller) module at left. To right is the cylindrical high-pressure fuel turbopump (HPFT). *Author*

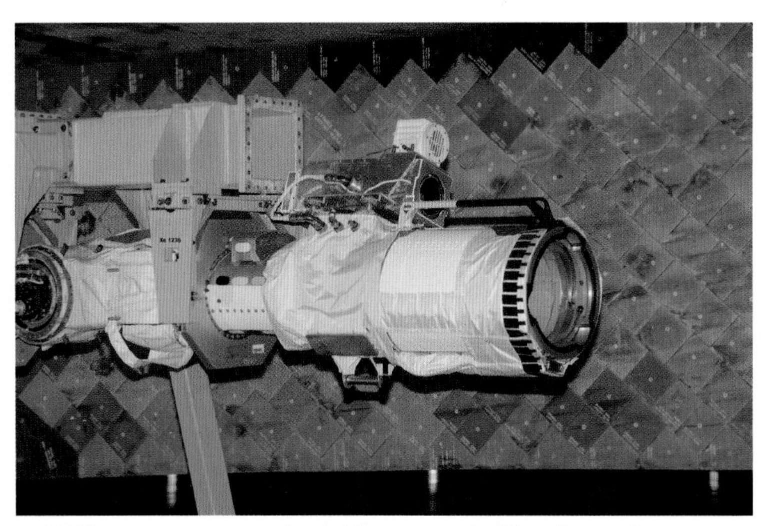

An RMS arm next to space shuttle *Discovery* at the Hazy Center. Components such as the end effector are shown. It used three cables to surround and center grapple fixtures on items equipped with them. *Author*

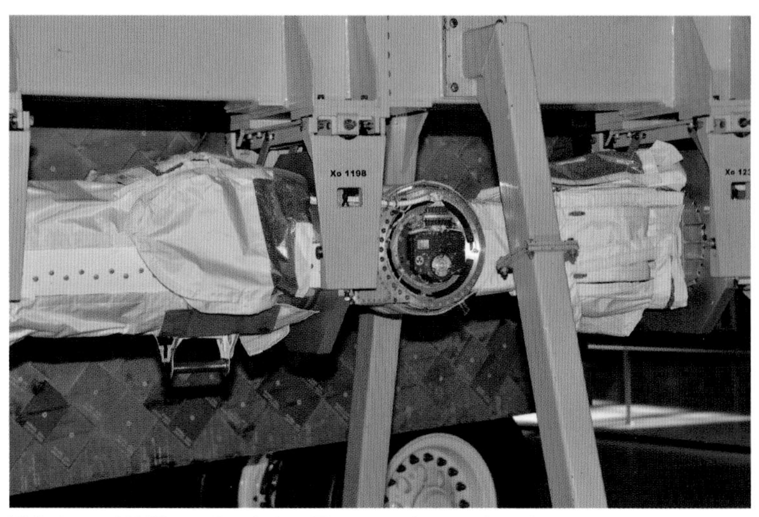

This is an RMS motion joint. They are located and move much as on a human elbow and wrist. The arm is operated by a payload specialist in the crew cabin, using a joystick while viewing RMS-dedicated TV monitors. *Author*

This angle of the end effector shows two items above the device: a TV camera and, above this, a light for use in shadow or on the dark side of Earth. Arm is 50.25 feet long and 15 inches in diameter. *Author*

RMS joint brakes. These are used for holding objects in place for extended periods. The arm has moved objects as heavy as the ZARYA ISS module, which weighed 42,637 pounds. The two-piece composite arm itself weighs only 93 pounds. *Author*

SRTM (Shuttle Radar Topography Mission) payload flew on STS-99. C- and X-band antennas mapped the earth in 3-D format, over 222 hours of use. This is on display at the Hazy Center with *Discovery*. *Author*

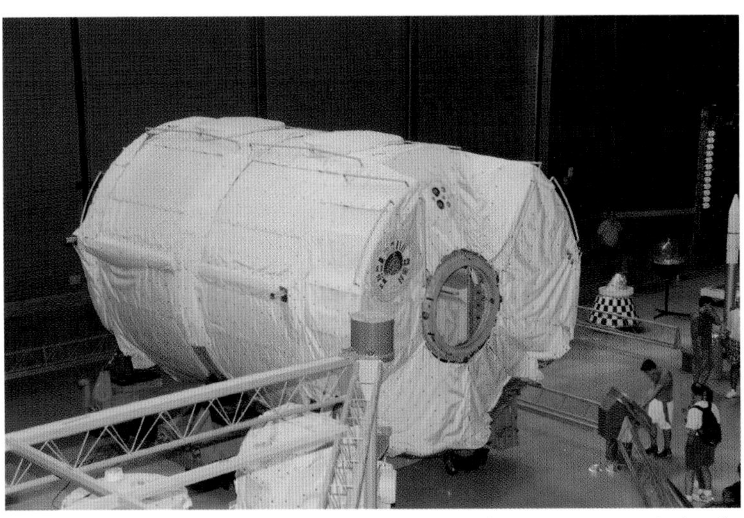

Spacelab module was a payload bay cargo flown on several shuttle missions to operate dozens of scientific and research experiments, with the orbiter providing the power and cooling needed. *Author*

The SPARTAN (shuttle pointed autonomous research tool for astronomy) free flier was deployed from the payload bay, then brought back after studies of the sun and solar wind concluded. *Hazy Center*, *author*

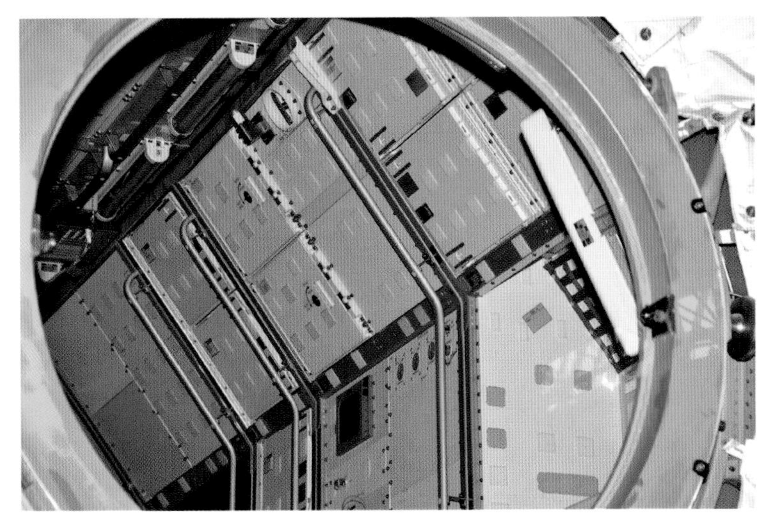

A look inside the Spacelab shows handrails, lighting, and lockers that, in this case, contain tools, maintenance equipment, and video systems. Blue ring connected to an access tunnel. *Author*

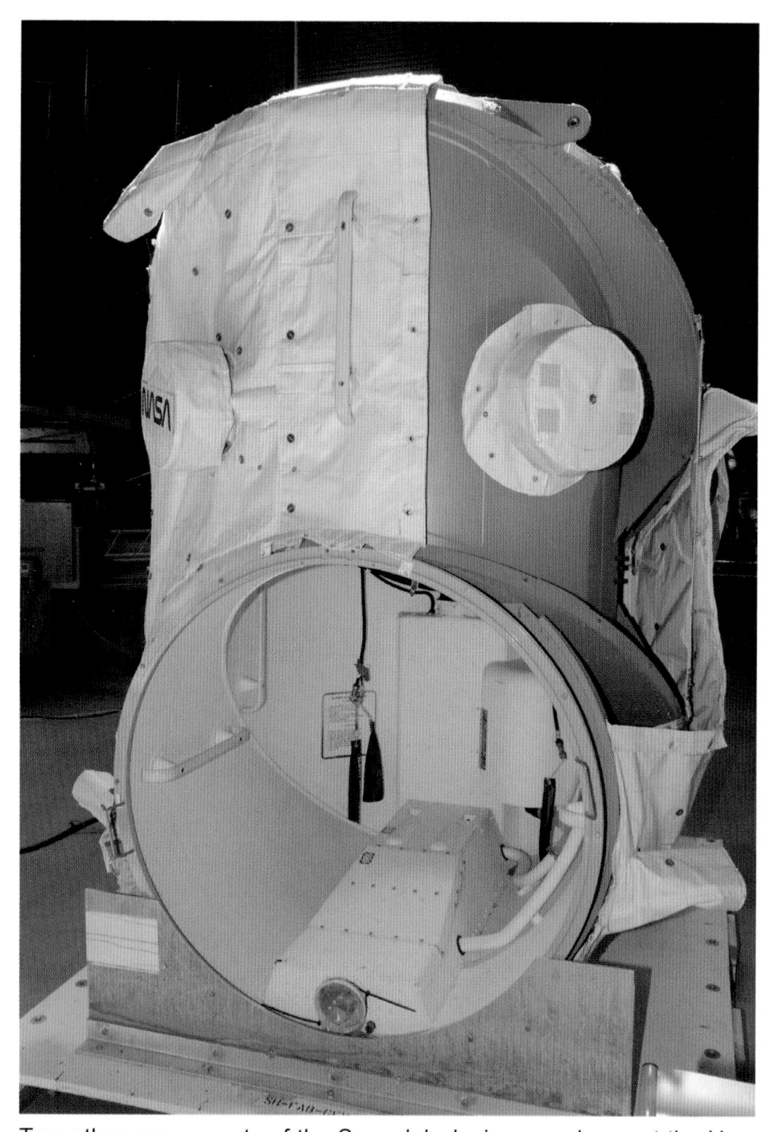

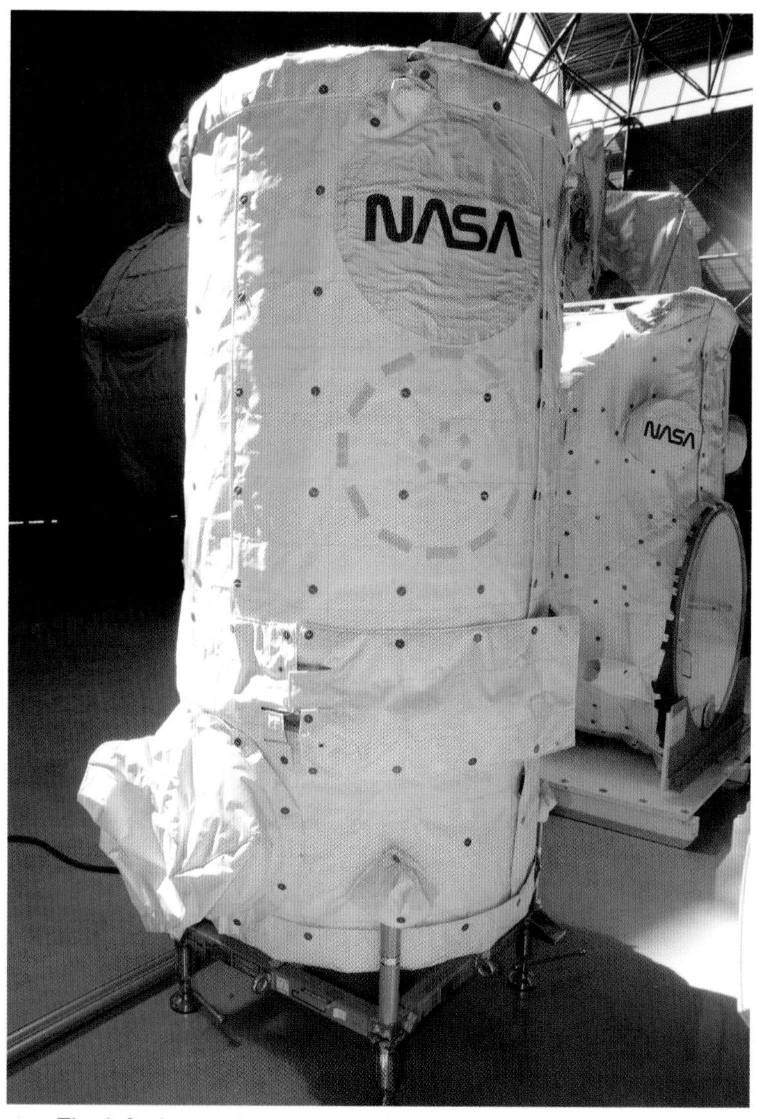

Two other components of the Spacelab design are shown at the Hazy Center. The left picture shows a part of the Spacelab transfer tunnel that shuttle astronauts and international partner crew members used to get from the crew cabin to the laboratory module. Hatch entrance view here is the section connected to the shuttle cabin. An elbow or "joggle" heading aft accounted for the height difference between the cabin/Spacelab entrances. Structure can be seen with the insulation removed. The "Igloo" in right image is one of two built that flew on shuttle missions where only pallet-mounted items were used. It provided power and supported instruments that flew on missions such as Astro-1, ATLAS-2, and ATLAS-3, among others. *Author*

At the Seattle Museum of Flight, the shuttle FFT (full fuselage trainer) is exhibited, with the added benefit of visitors being able to venture inside the payload bay. Both payload bay doors are open; the arm is stowed at upper left. At right corner is the black dish of the Ku-band antenna. Two aft-facing crew cabin windows are above a replica ODS. Looking through the access hatch can be seen some of the middeck stowage lockers. Yellow handrails are used during EVAs to move around the bay in a controlled fashion. *Author*

The USAF Museum in Dayton has the shuttle CCT-1 (crew compartment trainer). Astronauts got familiarized with orbiter internal layout and where everything was located in the crew module. Grey oval shapes on nose show RCS thruster locations. *Author*

A view of the CCT-1 middeck. Seat is shown, which is stowed after reaching orbit. One locker is partially opened. They contained any number of items such as clothing and personal gear. There are forty-two lockers, measuring 11 × 18 × 21 inches. *Author*

Looking forward from the aft station is the flight deck, with two mannequins in spacesuits. The instrument panel has the upgraded MEDS (multifunction electronic display system) EFIS or "glass" cockpit layout. *Author*

In the CCT-1 payload bay is this payload, known as TEAL RUBY or P80-1. It was a DoD device to be flown for exploring use of sensors in orbit to track ships and aircraft. The telescope unit is at left. It never flew due to cost and technology concerns. *Author*

The Huntsville Space & Rocket Center has this impressive full stack of the shuttle orbiter named *Pathfinder*, ET, and SRBs in an outside setting. The entire display is a replica. It does, however, give guests the ability to walk around it and gain insights as to the size of an entire assembly and the components that made up a shuttle in launch configuration. *Photograph taken by the author, October 2016*

The Space & Rocket Center also has a space camp facility that trains participants in aspects of life aboard the ISS. You can learn how to conduct experiments, do EVAs, work as a team, and operate the robotic arm. *Author*

Space camp also includes ISS events such as shuttle visits and the orbiter docking to the station and departing. A mockup *Atlantis* crew cabin simulates this not only for youngsters; adult space camp events occur as well. *Author*

ISS sections and the Kibo (Hope) Japanese experiment module are seen inside the space camp area. Those in the program conduct EVA tasks to retrieve as well as deploy items. They wear spacesuits and use tools unique to the ISS. *Author*

A space camp mission control center monitors ISS daily activities, crew members' taskings, and team duties. They check on the health of the station and advise those on board of upcoming mission goals. *Author*

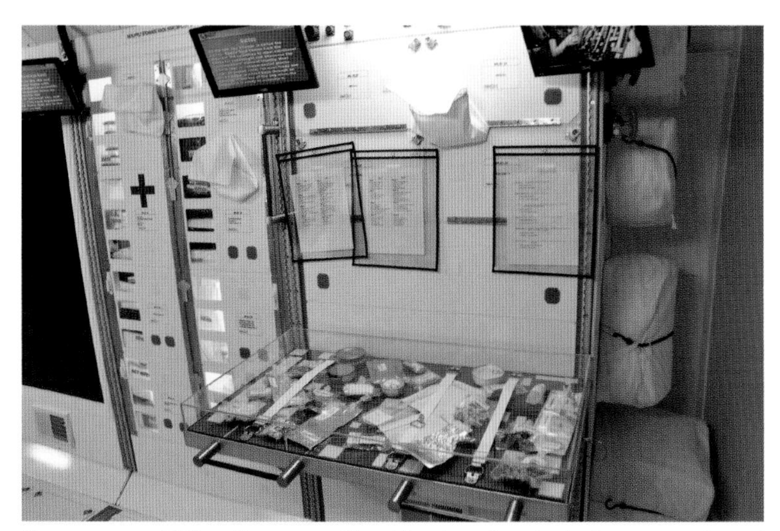

Space camp ISS galley area. Food is stored, selected, prepared, and made ready to consume. Great care is taken not to have items float away, since it can cause a potential safety hazard. *Author*

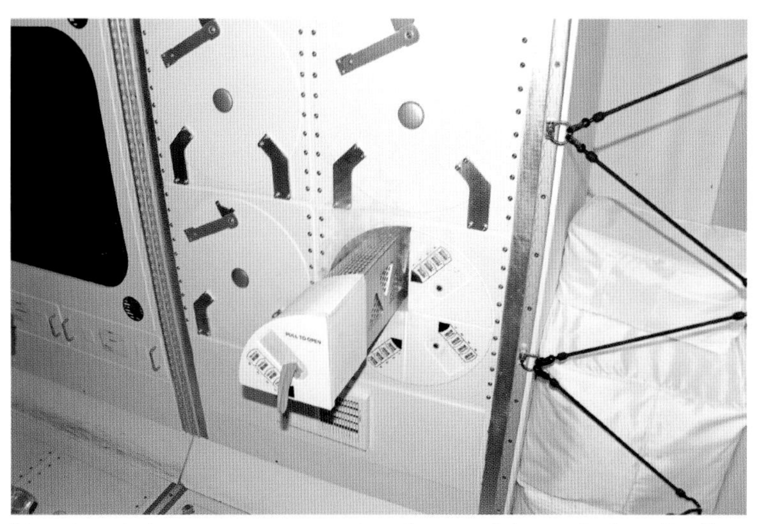

Experiments in some cases are in a frozen state, and when the time comes for them to be used, they can be extracted. ISS power generation provides for long-term freezing of many samples of scientific value. *Author*

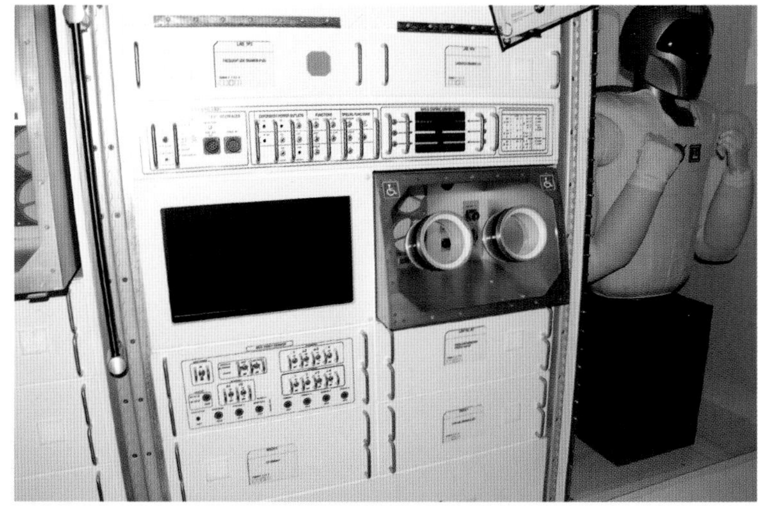

This is a part of the medical equipment, including video monitoring of in-progress events and recording for later analysis, and a station for handling materials within an enclosed space. *Author*

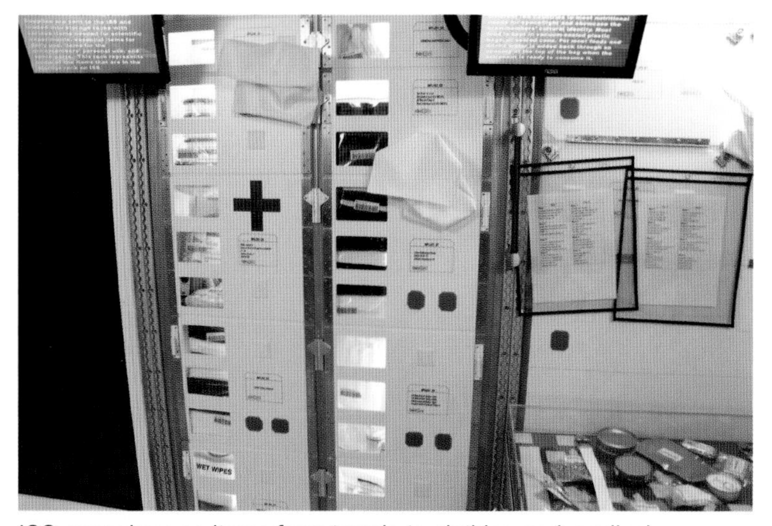

ISS everyday-use items from towels to clothing and medical gear can be accessed at these storage lockers. These are restocked after the arrival of a resupply vehicle from Earth. *Author*

ISS lavatory compartment. Urine as well as gray water and excess humidity in the air is recycled into potable water. This reduces dependence on outside sources for water and allows for continuous scientific work on board. *Author*

At this station the remote arm outside can be operated. Joysticks control movements, and monitors give three views from cameras strategically placed on arm sections. Crewmen can ride the arm by standing on a foot restraint. *Author*

ISS sleep station. Bungee cords secure equipment and the crew member; a laptop, air circulation, and lighting are all included. Crews work in shifts so that constant attention is given to the many experiments on board. *Author*

One of the favorite places to observe the earth and unwind is at the cupola section. It is also one of the prime places to take photographs, due to the wide perspective one gets when at this vantage point. *Author*

In June 1999, during a tour of the ISS processing facility, components of the station were in the midst of being assembled for eventual shuttle delivery. A consortium of nations are involved in ISS. *Author*

The *Leonardo* Permanent Multi-Purpose Module in the ISS processing facility being prepared. It flew in February 2011 on STS-133 and was installed in March. *Author*

A shuttle pallet as it goes through the steps necessary to set it up for an ISS mission. Much equipment came in to be inventoried, installed/tested, and then loaded on shuttle orbiters to be brought to the ISS. *Author*

One end section of *Leonardo* is to the right as it is surrounded by several work platforms and gear associated with this particular module. This was the last stop for ISS equipment before a flight. *Author*

Chapter 7

ARES I-X, ORION, STARLINER, DRAGON CARGO VEHICLE, AND SPACE LAUNCH SYSTEM

ARES I-X was a one-time flight of components in 2009 for the Constellation Program, which was canceled in favor of commercial flights to the ISS. It would have carried the Orion capsule to the moon as well as to the ISS. Despite this, some hardware has been saved, to be seen by visitors at the Space & Rocket Center in Huntsville, Alabama. The Orion capsule flew in 2014 and awaits a flight on the Space Launch System, or SLS. The CST-100 Starliner flew for the first time in December 2019. During its orbital flight test, or OFT-1 event, it made a landing in New Mexico. Another test flight (OFT-2) is scheduled to occur in March 2021.

SpaceX is busy with cargo flights to the ISS, manned launches, orbiting satellites for customers, and deploying its Starlink constellation of internet access satellites. It has developed the capability to recover and reuse items such as the payload fairings and first stages. These rockets make dramatic landings on floating barges at sea (or on land) without the use of parachutes, returning minutes after a successful launch. The NASA Space Launch System is in the final phases of checkout and assembly, for a tentatively scheduled test flight of the unmanned Artemis I mission possibly late in 2022.

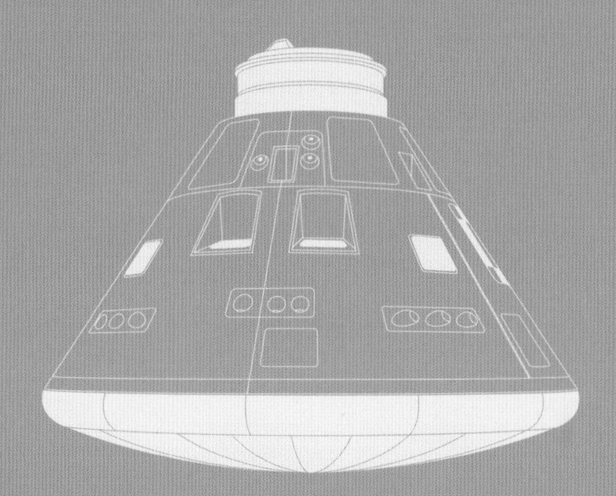

On October 28, 2009, the ARES I-X launch occurred, using a four-segment SRB first stage and a dummy upper-stage simulator. The flight lasted two minutes and cost $445 million. This was its one and only flight. *Author*

The RoCS (roll control system) rolled ARES I-X by 90 degrees after it lifted off. It provided a constant roll attitude until the stages separated. Propellants used were monomethyl hydrazine and nitrogen tetroxide. *Author*

ARES I-X J-2X engine that would have powered the upper-stage section. It stands 15.5 feet tall, with a 10-foot-diameter nozzle. Weight was 5,300 pounds and provided thrust of 294,000 pounds. *Author*

The two RoCS roll control engines were left over from the MX Peacekeeper ICBM missile program. They were externally mounted on the rocket body upper-stage simulator. This and the J-2X engine are shown at the Space & Rocket Center. *Author*

Orion capsule model at KSC in 2013. In most respects, it looks like nothing more than an Apollo-era command module of increased size. Thruster arrays and docking device are changed from the original. *Author*

At KSC is the Orion Exploration Flight Test 1 (EFT-1) capsule that flew on December 5, 2014. The first item of note are that it is tiled much as the Shuttle orbiters were. *Author*

Orion model at KSC in 2017 shows the latest configuration with docking-window area, thrusters, and umbilical panel connection. A changed docking device and the reflective exterior surface are also of note. *Author*

EFT-1 docking windows. Tiles are numbered and cannot be swapped from one area to another, just as it was on the shuttle orbiters. Also seen are thrusters that have red plugs to keep out foreign objects. *Author*

In July 2017, KSC had on display an Orion launch abort system (LAS) tower. It has the same function as the ones on Mercury and Apollo: to pull the crew capsule free of a malfunctioning rocket. *Author*

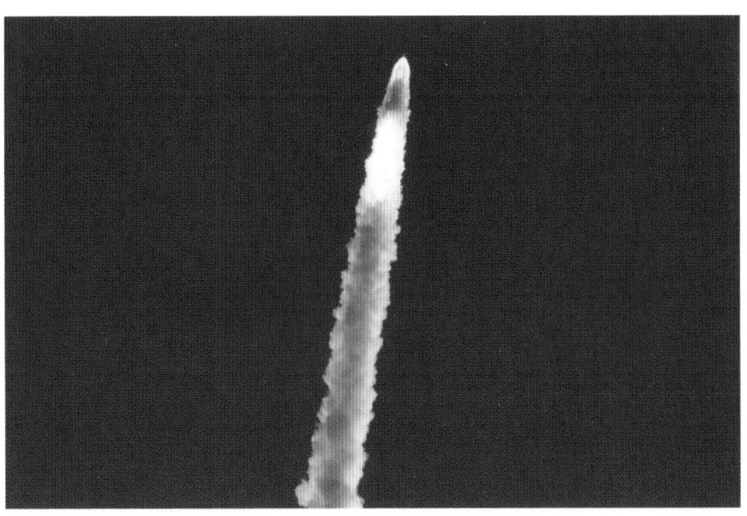

In this action image, the LAS has pulled the Orion capsule free and is propelling it away from the rocket at 600 miles per hour. LAS rocket motor fuel burns at a temperature of 4,456 degrees Fahrenheit. *Author*

On July 2, 2019, NASA conducted a live test of the LAS. The event was to ensure that the design and operation were sound and reliable. The Ascent Abort 2 test occurred at 7:00 a.m. and lasted three minutes, thirteen seconds. *Author*

Emerging from the exhaust cloud is the spent LAS, with the Orion capsule in free fall at lower right. The test was a success and proved the system was ready to perform its intended function, if needed. *Author*

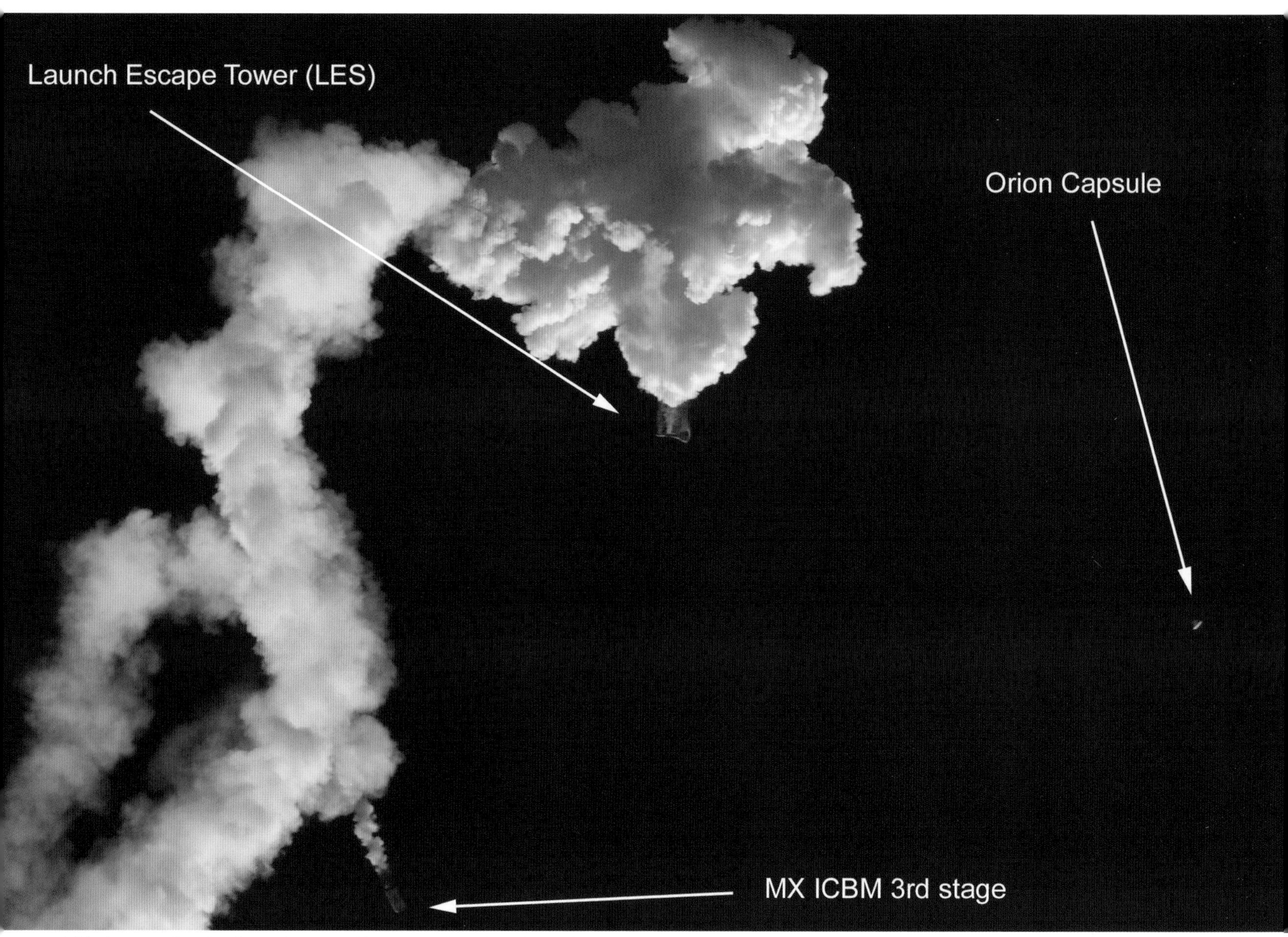

This image taken by the author standing on the beach at Cape Canaveral shows the conclusion of the Ascent Abort 2 test event. The rocket stage used was an MX Peacekeeper ICBM (intercontinental ballistic missile) third stage. The objects in the photo are at 31,000 feet on a clear morning. Everything visible in the scene fell into the ocean, sank, and was not recovered. The capsule ejected data recorders before impact with the water, which were retrieved for analysis. The space hardware used was expendable, proving that a safety feature did work as designed.

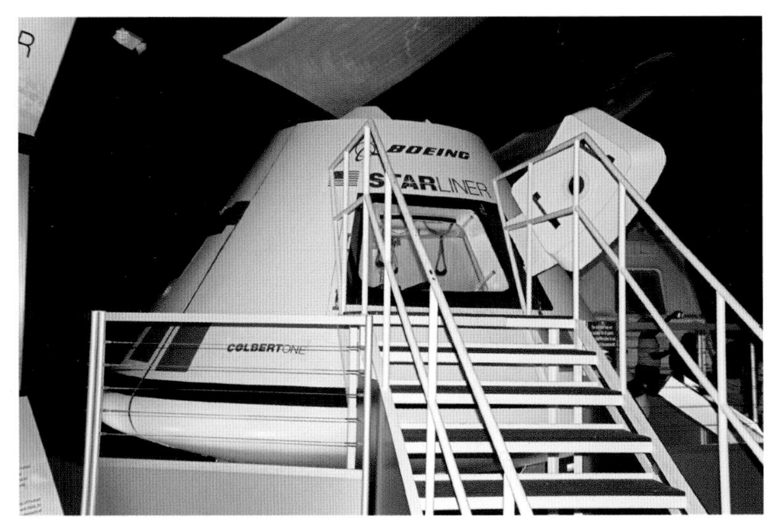

The CST-100 Starliner on display at KSC is a mockup not intended for flight. It is being designed to ferry crew or a crew/cargo mix to the ISS. It is a reusable capsule of up to ten flights and landings on the ground. *Author*

The CST-100 interior has, in this configuration, seating for a crew of five. Up to seven could be carried. Advances in electronics and glass displays have reduced weight and the spaces all the equipment formerly occupied. *Author*

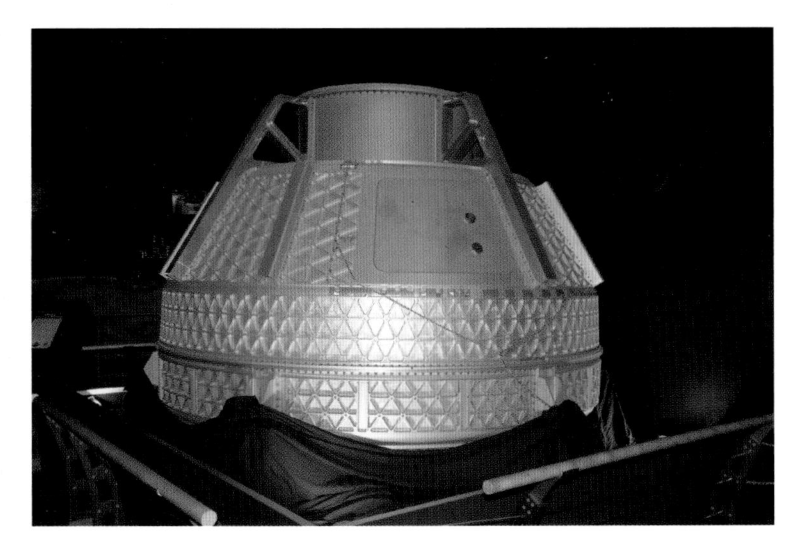

The CST-100 pressure vessel is of a weld-free assembly. Sections are of a honeycomb design composed of an aluminum alloy for reduced weight and high-strength qualities. *Author*

The KSC CST-100 exhibit also includes a SpaceX Dragon cargo vehicle that flew on an ISS resupply mission. In this photograph, designs and shape differences are seen between the two spacecraft. *Author*

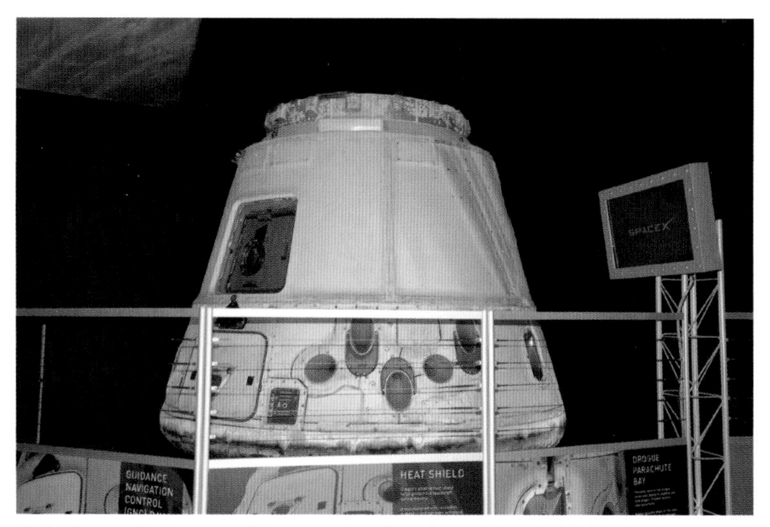

This SpaceX Dragon C2+ was the first American vehicle to dock at the ISS since the end of the space shuttle era; this event occurred in May 2012. Delivered were 1,567 pounds of food, water, clothing, and computer gear. *Author*

This panel was where an umbilical connection to external sources was located before launch. Once the connection was severed, internal components and power sources took over, verified by the ground before launch. *Author*

This access panel could be removed by ground crews to service or replace electronics associated with guidance, navigation, and control (GNC) equipment. Triangle bolt pattern was where a grapple fixture had been. *Author*

Seen on the Dragon cargo vehicle is an empty space where a drogue parachute had been stored. When deployed, the lines pulled free of the channel up to the nose section, where the main chutes were then opened. *Author*

On May 30, 2020, the SpaceX Crew Demo 2 launch took astronauts Bob Behnken and Doug Hurley in the Dragon capsule *Endeavour* to the ISS. The liftoff (*in the left picture*) was shot by the author from a location on the Cape Canaveral beaches south of KSC. A few days later, the first-stage rocket was towed into Port Canaveral. The rocket landed on a barge in the Atlantic Ocean, to be serviced and reused again on another mission. The stage has the NASA "Meatball" and "Worm" logos visible. Landing legs unfolded just before touchdown. *Author*

SpaceX Starlink satellites launch on January 6, 2020. A night launch is challenging to photograph, but the result is quite dramatic, recording the bright arc of flame against a black sky. *Author*

Go Navigator, one of the SpaceX fleet of oceangoing vessels, enters Port Canaveral on August 7, 2020, carrying the *Endeavour* capsule. It landed in the Gulf of Mexico days before. *Author*

Another Starlink launch, this one on November 24, 2020. A constellation of these satellites will eventually provide global internet access. Shot taken on the Cape Canaveral beaches with tripod; exposed for about ninety seconds. *Author*

The *Endeavour* capsule resting on board *Go Navigator* as she enters the Trident Turn Basin at Port Canaveral to perform safety checks of the propellant systems. *Author*

A look at the future of American space access is provided by this model of the Space Launch System (SLS) coming together at Kennedy Space Center for a possible launch later in 2022. This vehicle replaces the Constellation/ARES project. Height is 321 feet. Liftoff thrust is 8.8 million pounds, the most powerful rocket ever flown. Core stage has four RS-25 space shuttle–era engines and is 28 feet wide. SRBs are of five segments each, 177 feet long. Both the core stages and solid-fuel boosters are derivatives of space shuttle components. The Block I variant will lift 77 tons of payload into LEO (low earth orbit); Block IB will loft 116 tons. Block II will have the ability to place 143 tons into LEO. A moon mission will take the astronauts and cargo there in an Orion capsule in a single flight. Development costs in 2020 dollars amounts so far to $18.6 billion. Estimated cost per flight is $2 billion. *KSC photos taken by author in July 2019*

THE AUTHOR

John Gourley is a military veteran with a career as a crash/rescue fire protection specialist with the US Air Force, and as a civilian working for the US government. His lifelong hobby has been photographing military hardware, from fixed-wing aircraft and helicopters to warships, armored vehicles, and spacecraft. John contributes his photos regularly to fellow book authors and magazine and book publishers. He has documented many projects, from the building of new aircraft to the restoration of warbirds. He has also written several books on naval vessels. He lives in Newark, Delaware.

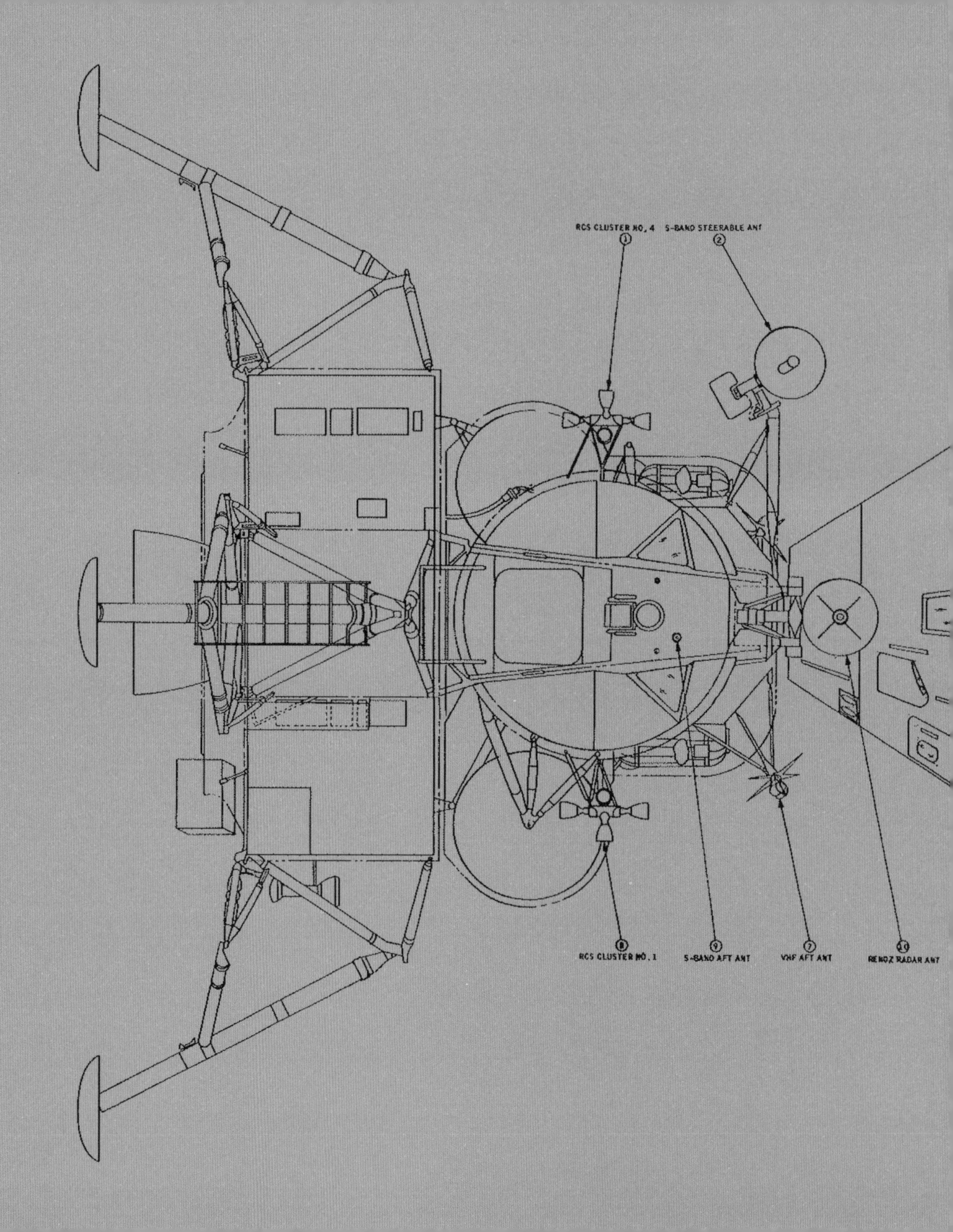
RCS CLUSTER NO. 4
1
S-BAND STEERABLE ANT
2
8
RCS CLUSTER NO. 1
9
S-BAND AFT ANT
7
VHF AFT ANT
10
RENDZ RADAR ANT